I0621545

Earlier Days Flying

Thomas U. McElmurry

Jung Works

Copyright © 2023 by Christine Jung.

All rights reserved.

No part of this work may be reproduced, stored in a retrieval system or transmitted in any form by any means, electronic, mechanical, photocopying, recording, or otherwise, without written permission of the publisher. For rights and permissions, please contact:

Jung Works

christine@jungworks.com

Book Cover and Illustrations by Christine Jung

1st Edition 2023

Contents

Preface   V

Editor's Note   VII

1. Pre-War Days   1

2. Army Duty   6

3. Light Bomber Duty   20

4. Off to Africa   24

5. Cuban Detour   31

6. Africa at Last   39

7. On to Italy   43

8. Back to the USA   66

9. You're in the Air Force Now   86

10. Korea   101

11. Holloman AFB   109

12. Test Pilot School   138

13. Working for the Navy   141

14. Sidewinder Testing   158

15. Aerospace Research Pilot School   173

16. Beyond the Air Force   188

17. Images   191

Review   204

Obituary   205

Bio   207

Found a Typo?   211

Also By Jung Works   212

About Jung Works   215

About Christine Jung   216

Acknowledgments   217

Preface

For some time, Tim has been encouraging me to write accounts of earlier times and experiences before I forget what I still remember about "Earlier Days." He even bought me a tape recorder and volunteered to transcribe the output. After listening to myself talk on the tapes, I switched to typing.

To work my way into this effort, I decided to start with something easy, "flying and airplanes." Hangar talk comes natural with aviators. The only significant problem that one has is holding down the tendency to embellish the stories. I have tried not to do that. However, some of the factual accounts of happenings are so "unusual" that the reader will probably have difficulty believing that they really occurred the way I've told them.

As I translated my recollections into written accounts of "happenings," I began to recall how often and how amazingly Almighty God and my Savior, Jesus Christ, have delivered me from failure, disaster, or both. In one of my favorite poems, Footprints, the author reveals her discovery of how many times Jesus *carried* her through the "impossible." She understood."

ISAIAH 40: 30-31
EVEN YOUTHS GROW TIRED AND WEARY AND
YOUNG MEN STUMBLE AND FALL;
BUT THOSE WHO HOPE IN THE LORD WILL RENEW
THEIR STRENGTH.
THEY WILL SOAR ON WINGS LIKE EAGLES.

The Bible clearly states that God has a purpose for every individual born into the world. Many of the details of that purpose are also clearly stated in the Bible. I would very much like to believe that God's purpose for my existence is, at least in part, served by my flying addiction. I am convinced that He had a hand in providing a path to a career in military aviation, a path around the seemingly insurmountable obstacle of being born totally deaf in my right ear.

My experiences and observations tell me that the reasons an individual is drawn to a job, profession, or endeavor are rarely clear. Role models could have been involved in my addiction to flying. If they were, I can't name them today. I do recall that, well before I became a teenager, I spent a sizable part of my "lawn mowing" money (25 cents for big lawns....10 cents for smaller ones) on flying magazines and balsa models. Just the sound of an airplane engine was enough to make me bolt from my seat at the dinner table and dash outside to watch the airplane until it disappeared.

June 3, 2003

Editor's Note

Walking into Thomas McElmurry's (or as I knew him, Papa Mac's) house, I was always first hit by the smell of *strong* coffee, then the biggest, heartiest hug and smile.

Papa Mac was a living legend, but I actually didn't find out about most of his stories until after he passed. He never talked about himself. When you were with him, *you* were the most important person in the world. He wanted to know everything about how you were doing and what was going on in your life.

Only once did I get to fly with Papa Mac. What an experience! He had a true gift for teaching, and I was fortunate enough to have experienced one lesson with him. Don't tell my mom, but he even let me take off and land. I still remember him directing me back to the field. I wondered how in the world he could pick out the tiny runway from way up high! But there it was. An adventure I'll never forget. I only regret I didn't get to do it again.

I remember Papa Mac had this pull up bar in his backyard. Even in his 80s, he was doing his age in sit-ups and pushups every day to stay in shape. One time, I showed him how many sit-ups I could do. He was impressed.

He had these games (flight simulation games, of course) on his computer since they were invented. He was never too old to learn. And never too old to stop having fun, having adventures, and investing in others. I recall him teaching a deaf student aerobatic flying lessons in his late years.

Back to not knowing much about his past when I was young, I do remember seeing an article with a framed metal in his office. He didn't want me to see it and read the article because it mentioned destroying a supply train during the war. The article implied horses were killed during the bombing, and he knew how much I loved horses and didn't want me to know any had died because of him. Another time, my brother and I were up in the attic looking for a board game to play and came across an entire truck of metals and commendations.

His garage had the typical tool chests and spare parts (all organized and labeled, as he liked everything.) But it also had someone's space suit, boxes of photographs (that I hope made it into some museum as many were of historical people and moments), books on physics and flight that were way beyond my comprehension… I even remember finding a guest book that they kept in their house decades before. NASA and USAF legends were in that book. I'm sure it's in a trash heap somewhere now.

If you search for "Thomas McElmurry" online, you'll read wonderful tributes from some of his former university students. You'll also find interesting and inspiring articles about his professional and personal life. He was passionate about flying – it's all he ever wanted to do. But he was also passionate about people. He was driven, genuine, adventurous (an adrenaline junkie, perhaps?), and generous.

I hope you enjoy reading these memoirs. I know he enjoyed writing them and living all the moments comprising these pages.

-Christine Jung
(Thomas' Granddaughter)

1

Pre-War Days

THE CURTIS ROBIN..... FIRST AIRPLANE RIDE!

CURTIS ROBIN

The exact date of my first airplane ride was lost to recollection a long time ago. It probably occurred sometime in 1937 or 1938. The details of the flight itself will always be remembered. It was probably the event that solidified teenage fascination and wishing into a lifetime addiction to flying. In the thirties, Alvin Guthrie owned and operated the Guthrie School of Aeronautics near Stillwater, Oklahoma. Occasionally, he flew his Curtis Robin from Stillwater to Batesville, Arkansas, to pick up a load of catfish. He always landed in a hayfield, which paralleled the White River viaduct south of Batesville. The fish were purchased from Mr. Porter Hargrove, a Batesville resident who had a fishing dock underneath the White River Bridge.

Guthrie almost always flew into Batesville late Friday and returned to Stillwater late Sunday. On Saturday and Sunday, he flew passengers at fifty cents per head. After the metal catfish tank was removed from the rear seat, the Curtis Robin had room for the pilot and three passengers. When my turn to fly came, I had the good fortune to be grouped with a guy and his girlfriend. Guthrie let the passengers decide on the character of the flight (stunts.........or gentle maneuvers). The young man, wishing to appear "macho", was the spokesman for we three. He chose "stunts."

CURTIS ROBIN FLIGHT PATH BENEATH THE WHITE RIVER BRIDGE

Guthrie certainly obliged. The flight couldn't have been more than fifteen minutes long. But it was worth far more than fifty cents! After takeoff, Guthrie made a wide circling climb over Batesville, as we marveled at the tiny switch engine, miniature cars, and doll houses. He then flew to a point about one mile up the river from White River Bridge. There, he asked if we would like to fly under the bridge. My male fellow passenger had to maintain his macho image by opting for the trip under the bridge.

The altitude couldn't have been more than 1500 ft as Guthrie pulled the aircraft up into a stall and entered a spin. In my mind, I can still clearly see the ground spinning around as we came straight down. The spin recovery and pullout were completed at just enough altitude to level out over the river at about 100 ft. Guthrie then let down to about 5 feet above the water and went under the bridge between the center pier and the bridge abutment. Wing tips had a lateral clearance of about 50 feet on each side. The top of the wing was about 15 feet below the bottom of the bridge. After chandelling to the right over the trees along the riverbank, Guthrie cut the power and made a power off landing in the hay field. That bite by the Flying Bug had a lasting effect on my pilgrimage through this life!

THE HEATH PARASOL

In the Thirties, Batesville, Arkansas, had no airport. Some time around 1938, a small dirt field was graded and declared an airport. But its operational life was brief. Without access to a rental aircraft and a flight instructor, there was no opportunity to learn to fly in Batesville. So, I started looking for alternatives. One of the flying magazines I purchased advertised a publication called "Trade-a-Plane." I invested a quarter and received about three mimeographed sheets, which listed aircraft for sale or trade. One of the listings was an unfinished Heath Parasol (without an engine) for sale by a guy in Kansas. He was asking $75.00. I had almost that much saved from my $6.00 per week job as a stocker and handyman at a Kroger store. So, with a two-week salary advance and my savings, the purchase was made. The Heath Parasol was a high wing, open cockpit, single-place airplane with a landing speed of about 40 mph. It could get off of the ground with a 40 hp engine. Performance with a 65 hp engine would have been very good.

Fortunately, the war came along before my savings climbed to the price of an engine. I had purchased and memorized the contents of a set of "Casey Jones' Flying Lessons." The plan was to complete the airplane and fast taxi it up and down the hay field until I felt confident enough to take off. The first flight would almost certainly have been much more exciting than the Curtis Robin ride.

A couple of years after I left Batesville with the Arkansas National Guard, Dad and Mother gave away the uncompleted Heath Parasol. All that is left is a sketch and Materials List for the engine mount.

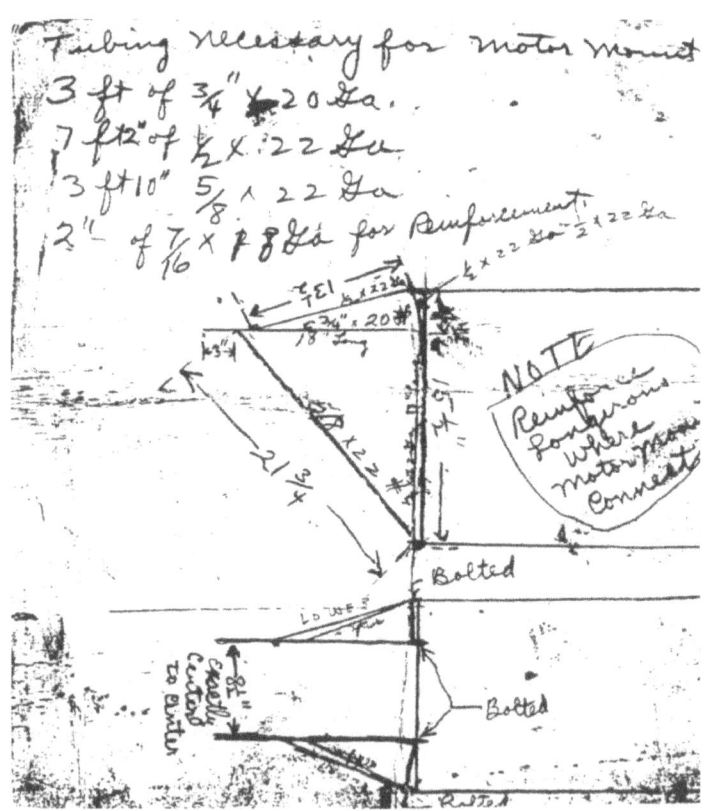

HEATH PARASOL

2
Army Duty

TRAIL TO THE ARMY AIR CORP

There was no reason to even imagine that joining the Arkansas National Guard in the spring of 1940 would ultimately lead to Army Air Corp Flying School in 1942; but it did.

Neither I nor my family had the funds for college when I graduated from high school in June 1939. This really didn't matter to me, for the college path didn't appear to lead to anything related to flying airplanes. I was well aware that two years of college were required to become an aviation cadet. I also knew that a deaf ear was absolutely disqualifying physically. So, after high school graduation, I worked in my dad's grocery store and in the Kroger grocery store that my brother managed. During this time, with the help of an outstanding physician with a big heart, I managed to squeeze by an Army physical. This allowed me to join the local Arkansas National Guard infantry unit, Company L, 3rd Battalion, and 153rd Infantry Regiment. From there, Divine intervention cleared a path to Army Air Corp Flight School.

THE PHYSICAL THAT OPENED THE DOOR!

In the Thirties, the Army offered an opportunity to seventeen-year-old high school graduates to earn a commission in the infantry as reserve second lieutenants. The program was called "Citizen's Military Training Camp, CMTC." Young men accepted into the program attended training camp each summer for four years. Those who satisfactorily completed the four years of training were graduated and commissioned as reserve second lieutenants.

I reported into Camp Robinson near Little Rock, Arkansas, in the summer of 1939 for my first encampment. Soon after arriving at Camp Robinson, I joined several hundred other "Basics" for an induction physical. After each man had been given a raincoat and a barracks bag, we were marched (out of step) to a series of long, one-story buildings. We were told to strip, put our civilian clothes in barracks bags, and line up for physical exams. The raincoats were to be worn when going from one building to the next. Doctors were spaced about every fifteen feet down the barracks. Each one checked one or more items on the list of things to be checked. The last physician in the series of examination stations checked hearing. I had no idea how I would get by with total deafness in one ear. But I figured there was nothing to lose by giving it a go. My heart sank, as I was greeted by the ear-doctor, when I arrived at his station. His question was, "You're deaf in your right ear, aren't you son?" All I could manage was a disappointed, "Yes Sir." He had observed me turn my head to hear questions as I progressed from one examination station to the next. I couldn't believe my good ear, when he smiled and waved me on with, "That's O.K." There's no way he could have known that his kindness would allow me to join the Arkansas National Guard, and later, to graduate from Army Air Corp Flying School.

CALL TO ACTIVE DUTY

The 153rd Infantry Regiment was Federalized on 23 December 1940 and ordered to active duty. In the spring of 1941, I became Platoon Sergeant of the 3rd Platoon of Company L, just in time to lead the platoon (on foot) over a large part of the Central Tennessee hills and lesser mountains during the Tennessee maneuvers. At that time, we had no platoon officers.

When the maneuvers were completed, the regiment was declared trained, and off we went to Alaska to prepare for the arrival of the Japanese. The fact that the Japanese surprised the troops at Pearl Harbor has always been a mystery to me.

A couple of months before 7 December 1941, we had our foxholes and pillboxes prepared on the beach at Seward Alaska. We manned them before daylight every morning. There was absolutely no uncertainty as to who we were expecting. I don't recall why; but, at that time, I had been reassigned as Sergeant of the 1st Platoon. We were deployed about a hundred yards from the edge of Resurrection Bay. In retrospect, I speculate on how many minutes we could have held the beach with one battalion equipped with rifles, machine guns, and mortars.

VIEW OF RESURRECTION BAY FROM BATTALION CAMPSITE

Everyone in the company knew that I would give everything I ever hoped to own to fly airplanes. It was, therefore, understandable that the Top Kick would immediately look me up when he received a memorandum announcing that enlisted men between the ages of 18 and 23 with a high school diploma could apply for admission to Army Air Corp Flight School. Successful applicants would become Aviation Students. The flight training would be the same as that for Aviation Cadets. But, instead of graduating as second lieutenants, Aviation Students would graduate as staff sergeant pilots. I smile when I recall that, even though I was man enough to be an infantry platoon sergeant, I had to have my father's permission to apply. I was 19 years old. My brother, my sister, and I persuaded Dad to sign the papers.

THE FLIGHT PHYSICAL

I rode train guard from Seward to Anchorage to take the physical, worried stiff about how to keep the physician from finding that the right ear was "nerve" deaf. As it turned out, the ear exam involved me walking to the far end of a long corridor and alternately holding a finger over one ear at a time while the physician whispered two or three numbers. Passing amounted to repeating the numbers correctly. I found it impossible to get the finger of the left hand closer than about an eighth of inch to the left ear opening.

The day after the physical, 7 December 1941, while I was waiting in the Anchorage Train Station to return to Seward, a civilian told me that the Japanese had just bombed Pearl Harbor. In February 1942, I arrived at Hancock College of Aeronautics, Santa Maria, California, for Army Air Corp Primary Flight Training.

PRIMARY FLIGHT SCHOOL: PT-13s AND PT-17s (STEARMAN)

Hancock College of Aeronautics had a grass field with a short asphalt runway paralleling the ramp on which the training aircraft were parked. The airfield was located just at the edge of Santa Maria, a pretty California town about twenty miles from the California coast. The cantonment area was not particularly impressive; and the grass takeoff and landing area was about like the Batesville hayfield. But the long line of yellow and blue biplanes parked on the ramp was the most magnificent collection of flying machines that I had ever seen. In my wildest daydreams, sitting in the cockpit of an unfinished Heath Parasol just two years earlier, I couldn't have imagined that someone would actually teach me to fly one of those.......for free!

The "floating on a cloud" euphoria was very quickly displaced by the sobering realization that only those fortunate few who could learn to fly well quickly would move on to Basic Flight Training. Over fifty percent of my class "washed out." Again, I suspect Divine intervention helped me over the hurdles. My first flight instructor, Mr. Portillo, was called to active duty as a second lieutenant, leaving me to fend for myself for a couple of weeks. Fortunately, I had just been cleared for solo flight when Mr. Portillo departed. I was allowed to continue flying solo until I could be assigned to another instructor. Eventually a Mr. Grey picked me up. Because of the break between instructors, my logbook shows only seventeen total hours of dual instruction in primary. One very early happening with Mr. Portillo is easy to remember.

THE GROUND LOOP

THE GROUND LOOP

I was having a good day on my second dual ride with Mr. Portillo. In fact, things were going so well that he decided to talk me completely through a landing flare, touchdown, and rollout. With only one flight of about forty total minutes of instruction in my logbook, my control inputs were completely mechanical responses to what he said. Still, things continued to go well. Had we broken off the exercise at the point where the aircraft made a flawlessly smooth three-point touchdown, we would both have been champs. Unfortunately, the challenge to take the exercise all the way through roll out was too much for Mr. Portillo. He continued to give rudder command inputs to correct a steadily diverging ground track. His decision to take control was too late. We skinned an outer wing tip all the way around a 360o ground loop. No serious damage was done to the aircraft. But a fine instructor suffered a great deal of embarrassment and fair amount of razzing by his peers.

PRIMARY ARMY AIR CORP FLYING SCHOOL
HANCOCK COLLEGE OF AERONAUTICS
SANTA MARIA, CALIFORNIA

BT-13
BASIC ARMY AIR CORPS FLYING SCHOOL
LEMORE, CALIFORNIA

THE GET ACQUAINTED RIDE

Mr. Grey was a bit on the stern side. But he was an excellent instructor. My first dual ride with him was structured like a check ride so that he could determine where in the flight-training program we needed to begin. Everything went well until he cut the power and asked for a forced landing. With great confidence, I selected and identified a landing spot. During the time without an instructor, I had perfected a dead stick landing system that worked every time. The technique was to always come in high on the approach to avoid undershooting. Then, lose the excess altitude when reaching the landing field was a sure thing. Nothing wrong with the idea...only the method used to lose the altitude needed some refinement. When the time came to lose the excess altitude, and before Mr. Grey could react, I sharply pulled up into a stall and immediately dumped the nose to recover. The altitude loss was exactly the right amount. Not then or at any other time did Mr. Grey ever raise or change the tone of his voice to me in the air or on the ground. With a calm, steady voice, he simply said, "Let me show you a better way to lose your excess altitude." To finish up the ride, he taught me the advantages of using forward slips.

VIEW OVER THE WING

Another of Mr. Grey's unforgettable lessons was taught without either of us ever acknowledging the fact. Still, I am certain that I was intentionally being taught a lesson, which I shall always remember. The instructor rode in the front cockpit of the Stearman. The front cockpit had a full set of flight instruments. The rear cockpit had the airspeed indicator face painted black, leaving the student with the engine instruments and an altimeter. Airspeed was determined by the sound the wind made passing through the flying wires. A very important item in the front cockpit was a mirror. The mirror was just above the instructor's head, so that he could observe the student any time he wished.

This particular flight included practice in spin recoveries. Spins were usually entered at an altitude of 2500 ft and recovery was initiated such that the rotation stopped precisely after the number of turns specified by the instructor had been completed. Very positive forward stick to stop the spin rotation was stressed by all instructors, particularly after a solo student in our class was killed when he failed to recover from a spin. If spin recovery practice was not the primary flight objective, Mr. Grey generally moved on to the next exercise after one or two satisfactory spins had been completed. Determined to impress Mr. Grey with my skill and adherence to accepted practice, I really put the muscle into the forward stick force at the recovery point in the first spin. As expected, the spin rotation stopped almost instantly. However, for about two seconds as I started applying back pressure to pull the nose up, I was looking over the top of the upper wing. Impossible! The eyeballs of a student in the Stearman's rear cockpit are at least a foot below the upper wing bottom. As I frantically tried to understand what happened, Mr. Grey's voice comes through the gosport tube, "Do another one." All during the climb back to 2500 ft, I mentally went through every detail of spin entries and recoveries. But no diagnostic light illuminated in my mind. Again, the stick went forward snappily at the second spin recovery point. The rotation stopped sharply.......and, again I'm looking over the upper wing top as I start the pullout! Now, I'm thinking, "This just has to be the last spin." However, before the thought is finished, Mr. Grey says, "Do another one."

As desperation was about to set in, Divine intervention reset the brain and the light suddenly dawned…"seat belt!" All through the climb, I held the stick with my knees and watched the instructor's mirror as I worked to fasten and adjust the seat belt without Mr. Grey discovering why the climb was somewhat erratic. There was no fourth spin. The view of the student in the instructor's mirror during the third spin recovery told Mr. Grey that his student had learned a valuable lesson that he would never forget.

I have never had the good fortune to again fly the Stearman since graduating from Primary Flying School.

BASIC FLIGHT SCHOOL: THE BT-13 (VULTEE VIBRA-TOR)

I was assigned to the Army airfield at Lemore, California, for Basic Flight Training. In 1942, Lemore was a comparatively small town located in the San Joaquin valley, just south of Fresno. It was a great location for a flying school. As I recall, we never lost a day of flying because of weather. Basic Flight Training repeated all of the things we did in Primary Flight Training and added several new dimensions to our flying adventure. We were introduced to instrument flying, day and night cross-countries, as well as day and night formation takeoffs and landings. Now, we flew in the front cockpit with a full set of instruments. The BT-13 had a two-position, variable pitch propeller... Low pitch for takeoff...high pitch for cruise. We had flaps that could be cranked down and up. I have forgotten how many turns were used for takeoff, and how many turns were used for landing. But I distinctly recall counting the turns. To we novice aviators, the BT-13 with its low wing, all metal construction, and sliding canopies was only about two notches below a combat aircraft.

THE ROLL ON FINAL

The only out of the ordinary happening that I recall from basic flight training was an unplanned roll on final approach in full view of a flight instructor. He was observing our landings from the ground. Brisk (steep) ninety degree turns to final, as opposed to shallow, flat (sneak up on the final) turns, were being stressed at the time. As always, my goal was to do 110%. I don't recall what my final turn airspeed was. But in looking back, I'm confident it was less than it should have been. At the final turn point, I rapidly put in full aileron with matching rudder. When the bank was about 85 degrees, I reversed the controls to stop the roll. The angular momentum of the aircraft was too much for the aerodynamic forces opposing the roll. I found myself on final approach upside down. The corrective action was obvious. I simply lowered the nose and continued the roll until I was right side up and made the landing. A great deal was said about the performance. None of the remarks were complimentary. I was overjoyed that the happening didn't result in a check ride.

GLIDER OPTION

About halfway through basic training, the Aviation Students were given the opportunity to transfer out of Army Air Corp Pilot Training into Glider Pilot Training. The inducement for making the transfer was a commission as a second lieutenant. Several transferred. The majority of us quickly decided that we had rather fly powered aircraft as sergeant pilots. Reading about the tough times the glider flight crews had during the Normandy invasion; it would be interesting to know how many of the Aviation Students from Lemore were involved.

Those of us who graduated from Basic in July 1942 moved further north in the San Joaquin Valley to Stockton Army Air Field for Advanced Pilot Training. My next BT-13 flight would be in October 1944.

ADVANCED FLIGHT SCHOOL: THE AT-6

Advanced Plight Training covered about the same things that we covered in Basic, but in an aircraft with a retractable gear. The AT-6 is a fun aircraft to fly. It does acrobatics well. Like a Pitts, it gains about 100 ft of altitude in a snap roll and can be safely snapped on takeoff. Snapping the aircraft on takeoff was not part of the training program.

ADVANCED FLIGHT SCHOOL: From Left to right: T. U. McElmurry, N. V. Meeks,
D. H. McFarland, Lt. B. E. Long (Instructor), D. E. McIver, E. L. McClure

Name Thomas U. McElmurry

Date of Birth Dec. 28, 1921.

Weight 160

Height 5 ft. 9 in.

Color Eyes Brown

Color Hair Brown

Date of Issue September 29, 1942

Signature T.U. McElmurry

Staff Sergeant

Ratings Held Pilot.

IDENTIFICATION CARD ISSUED GRADUATION DAY

DRESSED AND EQUIPPED
FOR A SOLO TRAINING FLIGHT

BLACKOUT NIGHT LANDINGS

Advanced introduced one interesting night flying exercise that we had not seen before. Someone with imagination decided that we should be prepared to land in blackout conditions. There were to be no landing lights of any kind on the airplane or on the ground, except two or three widely spaced flare pots placed at each end of the field. We were, of course, required to use navigation lights to avoid running into each other.

The schedule for the flight I was in fell on a moonless night. So, we had the challenge of landing in pitch-black conditions without a clue about how high we were above the ground. I don't blame my instructor for not attempting a demonstration landing. A landing from the back seat under blackout conditions would have been close to making the touchdown while under the hood.

As it turned out, I flew the aircraft into the ground twice, each time bouncing many feet into the air and holding the second descent to some 200 ft/sec to 400 ft/sec rate of sink with partial throttle for the second impact. After I finally found the ground during the second landing disaster, the instructor declared me likely to survive, got out of the airplane, and sent me into the black solo. I always knew that he was a fine instructor. That night convinced me that he also had a lot of common sense.

My next flight in an AT-6 would be made with a reserve outfit at Birmingham, Alabama Municipal Airport in 1946.

3

Light Bomber Duty

46TH LIGHT BOMB GROUP DUTY: THE A-20 AND THE DB-7

On 29 September 1942, the Aviation Students of Class 42-1 graduated as Sergeant Pilots at Stockton, California. Twenty-eight of us were sent by train to an Army Assignment Center at Salt Lake City, Utah. After a short stay there, we were reassigned to the 46th Light Bomb Group. At that time, the 46th was based at a desert airfield about five miles from Blythe, California.

When we reported in to the 46th, ten of us were assigned to the 87th Squadron, and were told to report immediately to the Squadron Commander. We never made it into the Squadron Orderly Room. Capt. Harvey Hogan, the Squadron Commander, intercepted us outside. After ordering us to fall in, he kept us in a brace while he delivered an ultimatum covering all of the undesirable things that would happen to any and all individuals in the group who even thought about getting out of line. After a group response of "Yes Sir" to his question, "Do you understand me?", we were dismissed to go to our quarters. We shared a one-story tarpaper barracks with the motor pool vehicle drivers.

Completely at a loss to understand why we were given such a reception, we talked with the motor pool personnel. Quickly we discovered that just before our arrival, a sergeant pilot had slow rolled an A-20 over the runway. Capt. Hogan intended to nip that sort of activity in the bud. We sergeant pilots flew with the squadron for four months without scratching an airplane, while two officer pilots killed themselves and their crews and a third totaled an A-20. Capt. Hogan called the sergeant pilots together and apologized for the reception. He also recommended us for 2nd lieutenant commissions. The recommendation was disapproved. But we very much appreciated the thought.

The 46th Light Bomb Group was a Replacement Training Unit, RTU. They took pilots from flying school and qualified them in the A-20 for ground gunnery, low level bombing, and the other combat flying skills. Actually, we flew A-20Cs and DB-7Bs. They were essentially the same aircraft. The DB-7Bs were the earlier models, which were sold to the French as dive bombers. The A-20Cs were primarily used by the British. In North Africa, the 47th Light Bomb Group was equipped with A-20Bs. I don't know what A-20 models the American outfits used in Europe and in the South Pacific.

The DB-7Bs had a 61 ft 3 in wingspan. The wing area was 464.8 ft^2. Empty weight was 11,400 lbs. Gross weight was 19,040 lbs. Maximum speed at 13,000 ft was 295 mph. Fuel capacity was 270 US gal. The power plant was two Pratt & Whitney R-1830-S3C4-G, 1200 hp engines.

The 46th group moved from Blythe, California to Will Rogers Field at Oklahoma City before we "New Heads" did any flying. Everything was much better at Will Rogers Field. We sergeant pilots had our own barracks. The 87th Squadron had its own hanger, and Oklahoma City was a great town! The favorite watering hole was the Silver Lounge in downtown Oklahoma City.

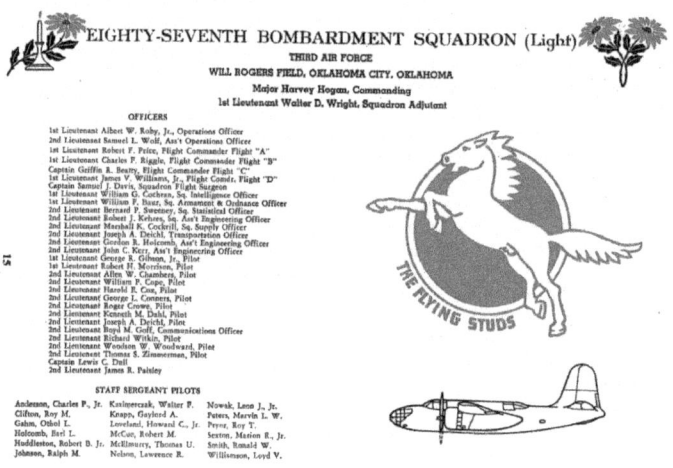

THE A-20 CHECKOUT

Flying the A-20 was outstanding. For a short time, I was a bit disappointed that I would be flying a twin-engine attack aircraft, and not a fighter. But the checkout flight gave me an entirely new perspective. The A-20 was several mph faster than the P-40s based at Will Rogers Field, and it was very agile for its size. Since the A-20 is a single pilot aircraft, there could be no pre-solo ride. Our preparation for the initial flight consisted of:

- Sitting blindfolded in the cockpit and touching instruments and controls as they were called out by the instructor.

- A briefing on what to do if we lost an engine.

- Instruction on how to use the brakes without overheating them. (As it turned out, using the brakes properly was a greater challenge than flying the airplane.)

Since none of us had ever flown a twin-engine airplane, going from an AT-6 to an A-20 was, to us, very impressive.

TRAINING FOR NORTH AFRICA IN A-20s AND DB-7s

The entire four-month combat flight training program was a joy. At least eighty percent of the flying we did was below 500 ft. Most of that was well below 500 ft. We did ground gunnery on the Salt Flats north of Oklahoma City. Aerial gunnery practice for the upper gunner and the lower gunner, who rode aft of the wing, was done over the Gulf south of Lake Charles, Louisiana.

The 87th squadron lost two A-20s and their crews during the training program. All that was found of one aircraft was a main landing gear strut with wheel. I found that part while buzzing the Gulf Coast beach south of Lake Charles looking for anything that might have washed up.

About halfway through our training program, the Army Air Corp created a "Flight Officer" rank. The difference between a Flight Officer and a Warrant Officer was the color of the bar worn on the blouse epaulets or shirt collar. The Flight Officer bar was colored blue with a gold stripe in the middle. All of the sergeant pilots in the 46th Light Bomb Group became Flight Officers.

4
Off to Africa

FROM OKLAHOMA CITY TO NORTH AFRICA

On 25 February 1943, about twenty flight crews of my group manned new A-20Cs and departed Oklahoma City for North Africa to help win the war. Each fight crew consisted of a pilot and two gunners. Billie Seamans and Jimmie Burns rounded out my flight crew.

A few days earlier, the first contingent of aircraft departing for North Africa really worked Will Rogers Field over as they were leaving. They simultaneously buzzed the field from all directions going between rows of aircraft, hangers, and just about any other place that there was a space between two objects wide enough to get the A-20 through. Before we took off, pilots had to report to Group Headquarters and sign a statement that we would not buzz the airfield. Of course, we didn't. But personnel at many airfields between Oklahoma City and Rabat, North Africa watched propellers blow dust as we made low passes arriving and departing. All of this immature behavior was not without cost. A pilot named Edens and his gunners were killed at Savannah, Georgia, when he flew through a control tower. Fortunately, the tower was unmanned. The tower personnel had just moved to a new control tower a few days earlier. As we were parking our aircraft after landing at Homestead Army Air Field, I pulled in beside an A-20 which had approximately eight inches of the propeller blade tips on both engines bent back ninety degrees. The pilot was not aware that he had hit the ground while buzzing.

Two other A-20s blew up en route. The first one killed everyone on board. Fortunately, only the pilot was killed in the second one. Just before the second one blew up, the gunners called the pilot on the intercom to inform him that they were standing in a pool of gasoline.

The pilot told them to immediately bail out, which they did. A few seconds later, the aircraft exploded. It turned out that a gooseneck hose which passed over the top of the engine to connect the outer wing fuel tank to the inner wing fuel tank was becoming disconnected. As a result, fuel was flowing down the wing into the engine compartment and fuselage. Eventually, some ignition source triggered the bomb.

THE DITCHING

The 46th Light Bomb Group A-20s, which left Oklahoma City together, parted company at Homestead Army Air Field, Florida. The overwater legs from Florida to North Africa were too long to navigate by dead reckoning. A navigator trained in celestial navigation was required. Since none of the A-20s had a navigator in their crew, we had to wait until some aircraft which had a cruise speed reasonably close to that of an A-20 came through Homestead with a navigator on board. A C-47 was much too slow to be the lead aircraft. B-25s and B-26s were fine. I was fortunate. When our turn came, an A-20B ferry crew with a navigator was passing through Homestead. On the morning of 16 March 1943, we left Homestead for Puerto Rico, arriving there in approximately five hours. It was a beautiful day for flying. The entire route had towering, scattered, puffy cumulus clouds floating over the white-caped water. My quarters at Borinquen Field were superb. I spent the night in a pre-war B.O.Q room for second lieutenants. In the thirties, second lieutenants lived well!

Next morning, 17 March 1943, was St. Patrick's Day. We took no special note of this fact. But the events of the day would make St. Patrick's Day special to me and my crew forevermore. This leg of the route had us landing in Belem, Brazil. The weather was forecast to be on the marginal side. About an hour from Puerto

Rico, we were in and out of broken clouds at about 5000 ft when I suddenly lost the left engine and was unable to maintain position in the formation. The rules of the game were that if you had a problem, and couldn't keep up, you were on your own. This made sense. Fuel margins for the long legs were small. Risking a formation of aircraft trying to save one was a bad gamble. Besides, the commander of the ferry aircraft was not actually responsible for us. We were simply hitch-hikers. So, after making several unsuccessful restart attempts, I feathered the left propeller and reversed course for Puerto Rico. At Savannah, Georgia, the aircraft had been loaded with tools and all sorts of cargo. The aircraft nose was filled up; and cargo was stowed in every available space aft of the wing. A 600-gallon fuel tank was installed in the bomb bay. This was full, except for the fuel that had been burned in the past hour. As loaded, the aircraft was losing between 300 ft/min and 400 ft/min on one engine. In an effort to improve our chances, I told the gunners, Billie Seamans and Jimmie Burns, by intercom to throw out as much as they could through the bottom hatch to decrease the aircraft weight. Before Billie could stop him, Jimmie immediately threw out the six-man life raft.

There was no question regarding what I was going to do. The pilot of an A-20 sits ahead of the wing and ahead of the propellers. As long as the aircraft is controllable and the surface below is acceptable for some sort of landing, a bailout attempt is the last choice. From my standpoint, I would either land the aircraft somewhere in Puerto Rico or I would ditch it.

The problem without a solution was what to do with Billie and Jimmie. The predicted behavior of an A-20 landing on water was that it would dive for the bottom. The normal exit for the gunners was through a hatch in the bottom of the fuselage. The odds of successfully exiting out of the bottom hatch of an A-20 diving in the ocean were very poor. Having them bail out, when we still had no land in sight and when no one even knew we were out there, was an equally poor choice. Once more Divine providence came to the rescue. When we were down to about 1500 ft, I saw in the distance a Coast Guard cutter approaching on a surface track from the right. I immediately told Billie and Jimmie to check their parachutes and be ready to bail out on my count as we passed over the Coast Guard cutter's course line. Later Billie told me that Burns was poised over the hatch when the call came to go. Billie helped Burns out with a push. When he saw Burn's chute open, he instantly followed. As it turned out, the Coast Guard cutter crew was watching the proceeding, and quickly picked the two up.

Shortly after the gunners bailed out, I caught sight of Aguidilla Point, a piece of land on the southwest tip of Puerto Rico. It was now clear that there would be a ditching. So, I got out of my parachute harness, jettisoned the top hatch over the cockpit, and unlatched my safety belt. As I was unlatching the safety belt, I thought, "You dummy, when you hit the water, there's nothing to keep you from slamming into the control column and the hardware in the front of the cockpit." But it was too late. The time to flare the airplane was now. So, after cutting the power on the good engine, I braced myself with my left forearm against the windscreen, and made the touchdown. My head experienced something close to a knockout blow when it contacted the windscreen. But the deluge of water pouring over the aircraft as it went down instantly revived me. In less than two seconds, I was frantically swimming for the surface. Popping out like a cork, I yanked the lanyards to inflate the Mae West, and heard a "woosh" of air as the gas from the CO_2 bottles escaped through the open inflation tubes. The next few minutes were spent inflating the Mae West with my lungs. My next thought was, "I wonder how long it will be before the sharks appear." A few tales in the Officer's Club the night before about shark attacks had me scanning the territory for anything that would minimize my shark bait time. As it turned out, a Puerto Rican fisherman in a small sailboat was already headed in my direction.

That day I learned something about sailing. I gave loud vocal directions for the fisherman to take a direct course rather than a zig-zag path to where I was. Of course, my vocal outburst was totally useless. The wind direction required that he tack. Besides, he didn't understand a word of English. In due time, the fisherman did get me into his skiff, safe and sound. Before we reached the beach, some Army Corp of Engineers troops, who had been conducting an exercise nearby, spotted us. They launched a powered boat and brought me the rest of the way to the shore. In an hour or so, the Coast Guard cutter with Seamans and Burns on board stopped about a block from the beach to determine if I had survived. That night we had a happy reunion at Borinquin Field.

POST-CRASH INVENTORY

JIMMIE BURNS (L) BILL SEAMANS (R)

A-20 RECOVER SOME TIME LATER

5
Cuban Detour

THE CUBAN DETOUR

O n 27 March 1943, ten days after the ditching, we departed Savannah, Georgia, for Homestead Army Air Field with a new A-20B. After a seven-day wait for an aircraft with a navigator, we once again left Homestead early on the morning of 3 April in a formation headed for Puerto Rico. The engine failure that led to the ditching had increased considerably the attention that I gave the engine instruments. About forty-five minutes out of Homestead, the rpm of one of the engines (I forget which one.) appeared to be behaving abnormally. At that time, the formation was passing within viewing distance of Nassau. My one foot wide "map roll"** indicated that the British had an airfield at Nassau. So, I peeled out of the formation and landed there.

> It was impractical to carry and try to manage in the cockpit all the maps covering the route from Florida to North Africa. So, I decided to make strip maps one foot wide and tape them together. These long strips were then rolled up so that they could be used like a scroll in the cockpit.

To my good fortune, the RAF mechanics based at the airfield were able to check the engine and give it a clean bill of health in less than an hour. Now I had a decision to make. Should I continue the flight to Puerto Rico without a navigator? Or should I fly back to Homestead. Having made the Puerto Rico leg once, and recalling that there were several island checkpoints along the way, I decided to go to Puerto Rico. Three hours later, everything was going great when Burns called me on the intercom and said, "I figured I had better let you know that fuel has been flowing out of the back of the left wing for quite a while." When asked, "How long is quite a while?" His response was, "It started shortly after we left Nassau." It took me less than a minute to discover what the problem was. I had left the fuselage fuel pump switch ON after starting a fuel a transfer to the right wing tanks. As long as the fuselage tank fuel-pump switch was ON, fuel would flow to the right wing tanks until they were full. Then, fuel would flow to the left wing tanks until they were full. If, after all wing tanks were full, the fuselage fuel-pump switch remained ON, the system would dump fuel overboard.

There was no fuel quantity indicator for the fuselage tank. So, there was no way of determining the remaining fuel total. Considerably more was required to get to Puerto Rico or to Homestead than the amount available in the wing tanks. So, unrolling my "map roll" again, and making a quick time and distance calculation, I determined that a southerly heading should take me to Cuba. Unfortunately, limiting the "map roll" width to one foot had left only a short piece of the northeastern tip of Cuba. This piece showed only two airfields. The most easterly airfield was very near the coastline. I decided this should be the most easily located. So, when Cuba came into view, I steered for the eastern tip. It turned out that the airfield was a very short dirt strip, much too short to make a successful landing. The map showed that the other airfield was near a town named Antilla, which was located on a bay. Hoping for the best, I penciled in a course line to Antilla.

I was elated to find that the Antilla airstrip was asphalt. I was not at all pleased with the runway length. Clearly, a fairly strong headwind was needed for an A-20 to stop in the stopping distance available. All the indicators (smoke, waving grass, etc.) showed a no-wind condition. Considering all the factors, I decided that I would touchdown as slow as possible, shutdown the engines at touchdown, and use maximum braking until I stopped. It was almost enough. The airplane was nearly stopped when it rolled off of the asphalt on to the rain-softened ground. The propellers had completely stopped rotating well before the runway end was reached. For a second, I thought we had made it. However, the aircraft had just enough momentum to overload two short drag braces on the nose gear. They broke. The nose gear folded backward until the aircraft nose rested on the mud. It would be almost two months before a B-18 with a ground crew from Camaguey, Cuba brought and installed two replacement drag braces.

The airfield at which we landed was less than a mile from the small town of Antillia. Antillia is located on Antilla Bay. At the time, Antilla Bay was a safe port for ships cruising the Caribbean. German submarines were very active in the area, and they were sinking quite a few cargo vessels. The United States Navy maintained a Liaison Office in Antilla to coordinate the Navy's involvement in protecting shipping. There was a small Cuban Army Military post in the town.

Pan American Airlines Clippers stopped at Antillia on their way to and from South America. Three or four times a week, the Cuban Airlines flew in and out of Antilla in a Ford Trimotor or a Lockheed Electra.

CUBAN DETOUR MOMENTS

RELOCATING TO MORE SOLID GROUND

EXTRACTING THE FLYING MACHINE FROM THE MUD

GETTING PRE-START ADVICE

AIRWORTHY AGAIN!!

DEPARTING FOR CAMAGUEY, CUBA

Several quality individuals made my stay in Antilla an unforgettable experience. Some of these provided invaluable help by solving problems for which I had no solution.

Mr. Arrue

Mr. Arrue was the Manager of the Pan American Airlines Station at Antilla. He had worked many years in the United States prior to taking over the Antilla Station. Relaying messages through the Pan American communication link with the States, he provided two-way contact with Homestead Air Force Base. Mr. Arrue's standing in the community gave access to any local support that was needed. He was in every way a generous, supportive gentleman.

Commander McGinnis

Commander McGinnis commanded the small Navy Detachment stationed at Antilla. He was particularly helpful in bargaining for the most cost-effective deal with the management of the Antilla Hotel where we both stayed.

United Fruit Company Management

I have lost the name of the United Fruit Company Manager who was responsible for the operation of the D-8 caterpillar that came to the Antillia airport and moved my A-20 to the airfield parking-ramp. There was no charge for this effort. They transported the caterpillar for over ten miles, moved the A-20 from its position in the mud to the ramp, and transported the caterpillar back home. Their contribution was a key part of getting the A-20 back to the States.

Cuban Army Soldier

Either Mr. Arrue or Commander McGinnis arranged for the Antilla Cuban Army Detachment to provide a guard for the A-20. I was at the airfield during the daylight hours for most days. I expected several soldiers to be assigned night guard-duty on a rotational basis. To my amazement, a single soldier was assigned to guard the aircraft for the entire two months that I was there. To my knowledge, he never left his post. He slept and did his hygiene things in the small building that served as a passenger terminal. His meals were brought to him.

Captain Terry

Waiting can get very monotonous, particularly when there are few people to talk to. The Cuban guard was a friendly troop. But he spoke no English and I spoke no Spanish. The Captain of the Ford Trimotor, which landed at Antilla a couple of times each week, Captain Terry, spoke fluent English. He welcomed the opportunity to chat with another aviator, as did I. I looked forward to his stops at Antilla.

Captain Terry was an interesting guy. It turned out that he had been an officer in the Cuban Air Force before Batista took over. Batista gave all of the officers the opportunity to move to his side of the revolution or be imprisoned. Captain Terry chose to join those who resisted the takeover. The resistance was short lived. Batista had done his homework well.

Army Air Corp Crew from Camaguey, Cuba

Shortly after we landed at Antilla, I learned that a U. S. Army Air Corp flying organization was stationed at Camaguey, Cuba. With help from Commander McGinnis, they were contacted for assistance. Since there was no reason to keep Seamans and Burns at Antilla, they dispatched a B-18 to Antilla to see what our situation was and to transport Seamans and Burns to Camaguey.

A few weeks later, when the nose gear parts arrived at Camaguey, a second visit was made with some aircraft mechanics to install the nose gear parts.

Cockshott Family

Antillia had a main street, which ran from the bottom of a hill at the north end of the town to the Antilla Bay shoreline at the south end of the town. There was a small circular observation building with a roof and no sides at the top of the hill.

One sunny day, while I was surveying the countryside from this hilltop, I noticed two young ladies making their way up the path to my location. One of the young ladies was obviously Cuban. The other's ancestral chain was clearly European.

She was the first to speak. When she introduced herself as June Cockshott, there was no mistaking her English heritage.

It turned out that June's companion was a friend who had traveled with her by train from a United Farm Plantation about twenty miles away. Their purpose for coming to Antilla was to meet the pilot of the A-20 which had landed at Antilla and invite him to visit the Cockshott family. June's father managed the plantation. Mr. Cockshott had been a WWI fighter pilot, flying Sopwith Pups. We instantly established common ground. Mrs. Cockshott was a gentile, warm English lady. She was always a joy to be around. June was a pretty, charming, talented young lady. Many enjoyable hours were shared with the Cockshotts while I was in Cuba. We kept our friendship active by corresponding throughout WWII and for many years thereafter. On 22 May 1943, I flew to Camaguey. The next morning Seamans, Burns, and I flew to Homestead Army Air Field.

6
Africa at Last

AFRICA AT LAST

The time estimate for completing the repairs on the damaged A-20 was many weeks. And a replacement A-20 was not available. My two gunners were reassigned as B-17 gunner replacements. My orders remained the same, except that I would be flying as a passenger on a C-54. When the C-54 landed at Natal, Brazil, for an overnight stay, I checked at operations to see if there was an A-20 on the airfield without a pilot to ferry it... there was.

On 9 June 1943, I followed another A-20 with a navigator from Natal to Accension Island, a piece of British real estate in the middle of the Atlantic Ocean with an airfield. The next day, the two-ship formation flew from Accension Island to Accra, Africa. At Accra, ground crews removed the bomb bay ferry tank from my aircraft. The mechanics discovered that a solenoid adjacent to the ferry tank had burned up somewhere between South America and Africa. They congratulated me on my good fortune. In their opinion, the potential for an explosion had been high. With God in charge, potentials don't mean a thing.

On 13 June 1943, equipped with full sized maps, I navigated my way from Accra to Dakar with a refueling stop at Roberts Field, Liberia. The Sahara was a thousand-mile leg with no more checkpoints than the Atlantic. Again, I had to wait for a lead ship with a navigator on the crew. The navigation task was made even more challenging by the 600-gallon reduction in fuel available. Without a fuselage tank, a refueling stop at Tindouf, a French Foreign Legion outpost in the middle of the Sahara, was a must.

The Sahara could be more treacherous than the ocean when sandstorms came up suddenly. Just a few weeks before I arrived at Dakar a flight of P-40s lost contact with a Lockheed Hudson lead aircraft in a sandstorm. Only two of six pilots survived. One pilot immediately turned due west when he became separated and flew until he found an inhabited spot in Spanish Morocco. There he bailed out and was rescued. The second survivor wandered around until he ran out of fuel and bailed out. His guardian angel had navigated him to a point directly over a tribe of nomads.

While waiting at Dakar for a lead aircraft, I was watching Martin B-26s take off for the flight across the Sahara when one lost the left engine as it was passing over the departure end of the runway at about a hundred feet. The aircraft immediately rolled inverted and exploded on ground contact. That afternoon, all of the officers waiting at Dakar were required to attend the crew's funeral at a nearby cemetery. On 21 June 1943, after refueling stops at Tindouf and Marrakech, I landed at Rabat, my destination. With all of the happenings between Oklahoma City and Rabat, the total transit time had been four months. The story about the two groups of A-20s that made the trip from Oklahoma City to North Africa in the Spring of 1943 became well circulated "hanger talk" during 1943-1944. A-20s blew up in the air. A-20s crashed buzzing. A-20s ditched in the ocean. We strewed A-20s from Oklahoma City to Casablanca. At least fifteen percent of those who started never arrived.

THE MONTH AT RABAT

Time passed slowly at Rabat waiting for assignment to an operational unit. The Afrika Corp troops who escaped from North Africa were being pushed out of Sicily by Patton and elements of the British Eighth Army. Army Air Corp units were in the process of relocating to airfields in Sicily. So, there was no hurry to get us assigned. For a month, we flew training missions out of the Rabat airfield and ferried replacement A-20Cs to British airfields.

FORMATION PRACTICE AT RABAT

On 5 July 1943, I was assigned an A-20C to deliver to the British at Setif, a British airfield between Algiers and Telergma. The trip should have taken about two days. But, when I landed at Algiers to refuel, I met some ferry pilots who were about to take some A-36s to Sicily. They were concerned about how they would get back to Algiers. As I listened to their conversation, it occurred to me that a trip to Sicily would certainly be more interesting than hurrying back to Rabat for some more waiting. So, I volunteered to follow them to Sicily and to bring them back to Algiers. As it turned out, Bob Hoover, the guy who would later become a famous acrobatic pilot and test pilot, was the A-36 flight leader. To show his appreciation for the lift home, he said that if I would stay overnight in Algiers after our return from Sicily, he would let me fly any fighter parked on the ferry flight line. Naturally, I stayed overnight.

The next morning, the choices narrowed to a P-38 and an A-36. Had I known that a few months later I would fly combat in A-36s, I would definitely have taken the P-38. Unfortunately, I chose the A-36. I never had another opportunity to fly a P-38. The choice of the A-36 wasn't all bad, however. When I landed from the A-36 flight, they told me that they were taking some A-36s to Tunis. And they were one pilot short. Sometimes, one is fortunate enough to be at the right place at the right time. I joined the flight to Tunis.

Finally, on 9 July, I delivered the A-20C to Setif. The British were not as appreciative as the Algiers ferry unit. In response to my query about transportation possibilities for return to Rabat, the British Operations Officer indifferently advised that how I returned to Rabat was entirely up to me. I suspect that at some point in his career he had crossed swords with an American. Someone other than the British Operations Officer showed me the road to Telergma. I started walking in that direction. Fortunately, after I had meandered down the road for about an hour, a couple of British soldiers in a jeep gave me a ride to the Telergma Airfield. Walking the distance would have taken days.

7

On to Italy

85TH LIGHT BOMB SQUADRON, 47TH LIGHT BOMB GROUP DUTY: A-20B

On 19 July 1943, with orders assigning two new gunners and myself to the 47th Light Bomb Group, I departed Rabat for Sicily in a new A-20B. The orders specified the wrong airfield in Sicily. No one at that airfield knew where the 47th was based. They suggested that I go to Aggrento and check there. It turned out that the 47th wasn't at that airfield. But they did know where they were. The third try on 20 July was successful. We found them at Communeli airfield on the west side of Sicily. The stay at Communeli was short, only about a week.

The group moved to a flat piece of ground about ten miles south of Catania, Sicily. The dirt landing-strip had been used by a German ME-109 outfit. In their haste to leave, they left behind about eight un-flyable ME-109s. We pleaded with the 85th Squadron C.O. to let us repair and fly one of them. But he wisely turned us down. A pilot of a P-40 outfit had just been killed flying one at a fighter field on the west side of Sicily. In addition, our maintenance personnel had their hands full keeping our A-20s operational. Takeoffs and landings in thick dust clouds didn't make their jobs any easier. To reduce the damage from dust, we took off and landed three abreast, allowing a minute or two between takeoffs and landings for the dust to partially clear. The 47th Light Bomb Group was a great outfit, as was the 85th Squadron to which I was assigned. The A-20B was a joy to fly. For its time, it was fast. The A-20 was very agile for its size. I looped it and slow rolled it as one would a fighter. We lost quite a few. But, compared to some other aircraft, it took battle damage well.

During my combat tour with the 47th, we moved several times to relocate to airfields nearer the bomb lines. Bomb lines were the lines of engagement between German and Allied ground troops. We attacked anything on the German side of the bomb line. We tried to never attack anything on the Allied side of the bomb line. Sometimes mistakes were made.

From Catania, Sicily we moved to a dirt runway west of Foggia, Italy by way of Taranto. The Group stayed at Taranto for a few weeks while the airfield west of Foggia was being prepared. Shortly after Christmas of 1943, we moved to a dirt field near the small town of Ottaviano at the base of Mount Vesuvius.

THE AIRFIELD NEAR CATANIA

CANTONMENT AREA

NEARBY ME-109 CRASH SITE
THE TERRAIN IS ROCK UNDERNEATH ABOUT 5 INCHES OF DIRT.
NOTHING LEFT BUT BITS AND PIECES.

THE GERMANS MADE SURE EVERYTHING WAS UNFLYABLE

THE GERMANS LEFT EIGHT ME-109S

TEMPORARY DUTY WITH AN AMERICAN BEAU-FIGHTER SQUADRON

When the time came to move operations from Sicily to Italy, the 47th relocated to Taranto, Italy to wait for a grass field west of Foggia to be readied. While the group was waiting at Taranto, a request came through for a few 47th pilots to go TDY to an American Beaufighter outfit stationed near Syracuse, Sicily. Each pilot would take an A-20 with him. German JU-88s flying out of Yugoslavia had been giving cargo ships sailing up the Aegean Sea a hard time around the clock. The plan was for the A-20s to provide protection during the day and Beaufighters to take care of the night hours. About six of us volunteered. I believe we spent about a month with the Beaufighter outfit.

Early in our temporary tour of duty, the Squadron Commander decided to allow some of us to check out in the Beaufighter. The Beaufighter had a deserved reputation for being a ground-looping s.o.b. So, an up-front mandatory step in the checkout sequence was to demonstrate a capability to maintain direction control during the takeoff roll and during the landing roll. None of us were ever checked out. I passed the acceleration to lift-off speed and post-abort roll out. But, a happening among the ranks of the regular Beaufighter pilots terminated the checkout program before any of the TDY troops could get airborne. By chance, I was standing on the side of the dirt runway observing the Beaufighter take off, when the aircraft ground looped so severely that an engine was torn off of its mount and went tumbling down the runway. The Squadron Commander called off the checkout program.

BRISTOL BEAUFIGHTER

47

FLYING THE SPITFIRE

There were times when day patrol duty for we 47th TDY pilots was slow. During one of these periods, I decided to hitchhike to an aircraft repair depot on the west side of Sicily and see if perhaps they could use the services of a volunteer fighter pilot. Technically, I was indeed a fighter pilot. I had two flights in an A-36, and broadminded folks could classify our patrol duty in A-20s as fighter time. My plan was to allude to some kind of P-40 experience (Some statement a little stronger than, "Familiarity with the cockpit," but less than, "Substantial flight time in the aircraft"). As good fortune would have it, there were no P-40s to be ferried. But they did have a Spitfire that needed ferrying to an American Spitfire outfit near Naples. With a suggestion from me that, "A P-40 pilot should have no difficulty flying a Spitfire." They opted to let me deliver the machine. The engine start was acceptable. But the taxi to the takeoff strip must have raised some misgivings in their minds about the decision they had made. For sure, the takeoff raised the hair on their scalps. It was terrible. Fortunately, by the time I arrived at the Beaufighter Strip I had a better feel for the light controls. The landing wasn't too bad.

For two or three days, I flew the Spitfire without anything being said. Perhaps, the management thought some visiting pilot had business there, which they didn't know about. Finally, one morning when all of the pilots were gathered in the mess tent for breakfast, the Squadron Commander asked, "Who does the Spitfire belong to?" My response, "It's mine Sir." sparked an immediate second question, "Where in the H......did you get a Spitfire; and what do you think you are going to do with it?" My account of the ferry plan brought an immediate "Veto" and an order to take it back where I got it.

My Spitfire V

MY SPITFIRE V

Taking the Spitfire back where it came from was the last thing I wanted it to do. Recalling that there was another repair depot at the Catania Airfield just north of Syracuse, I decided to drop it off there. That day, I flew the Spitfire to Catania, parked it in a spot near what appeared to be the engineering tent, walked into the tent, and announced to the person that appeared to be in charge, "I have parked your Spitfire in the revetment just outside." Without waiting for comments, I did a quick about face and departed.

A short time later, we flew our A-20s to the 47th's new home, a grass airfield a few miles west of Foggia, Italy.

MY QUARTERS AT THE FOGGIA AIRFIELD

The 8' X 10' building was made from scrap material obtained from bombed out Foggia Main Airfield, adjacent to the city of Foggia.

A-20 COMBAT OPERATIONS

Unlike the A-20 outfits in the South Pacific, which strafed and dropped parachute bombs at low level, we did our bombing in six ship formations from about 8000 ft altitude. The German 88s (anti-aircraft guns) were very effective against targets flying a straight course for longer than the time required for the 88 projectiles to travel from the gun muzzle to eight thousand feet. The travel time was on the order of five seconds. To make things more difficult for the German gun crews, when we suspected or knew that we were vulnerable to 88 fire, the pilot of the lead aircraft would continuously change altitude and heading as he took the formation in the general direction of the target. At the last minute, the lead aircraft would hold constant altitude and heading, making small heading adjustments in response to directions from the bombardier. (Only the formation leader had a bombardier on board.) The straight and level part of a bomb run was about a minute long. This was the time when most of the damage was done. The guessing game played by the German gun crews and the formation leader while the formation was maneuvering was roughly a fifty-fifty proposition. The gun crews were definitely in the driver's seat when the formation flew straight and level. The 88 gun crews would play games. Sometimes they would let the first squadron pass overhead without firing. Then, surprise the trailing squadrons with all batteries blazing away.

A crewman knew that flak was in the vicinity of the formation when it sounded like a small bomb exploding. When it sounded like a whip cracking, it was very near the aircraft you were in. As mentioned earlier, the A-20 was a rugged aircraft. I never had to rely on this fact during any of the missions I flew. I encountered nothing more serious than severed hydraulic lines and occasional holes in the aircraft skin during my tour with the 47th. Others were not so fortunate.

Losses came in a variety of ways. During a bombing run over Cassino, Otto Bey, one of my tent mates, slowly rolled out of the formation into a steep spiral. His two gunners bailed out. The aircraft continued in a steep spiral until it hit the ground. We never learned what happened. On a mission over Anzio Beachhead, the number two aircraft in my flight took a hit in the bomb bay and blew to bits. I was below and behind him in the formation. With my peripheral vision I saw the aircraft instantly disappear. One instant the A-20 was there. The next instant, it was gone. As I recall, the pilot already had his orders to return to the States. The outfit was short a pilot that day. And he volunteered to fly one more mission. It was the time for him and for his two gunners.

I completed my combat tour with the 47th Light Bomb Group while the outfit was at Ottaviano. Another of my tent mates, Terry Edmonds, and I transferred to the 86th Fighter Bomber Group to fly a second tour in A-36s. Shortly after we transferred, the 47th made an unscheduled move to the airfield at Naples. They left Ottaviano in a hurry to save the airplanes when Vesuvius erupted. Ashes were dropping on the Ottaviano airstrip.

Not until November 1944 did I have another opportunity to fly the A-20.

THE VESUVIUS ERUPTION WHICH CHASED THE 47TH LIGHT BOM GROUP OUT OF THE OTTAVIANO AIRFIELD

SERGEANT CONRAD, CREW CHIEF OF OUR A-36

DUTY WITH THE 86TH FIGHTER BOMBER GROUP: A-36 INVADER

527TH FIGHTER-BOMBER SQUADRON

Terry Edmonds and I started flying with the 527th Fighter Bomber Squadron of the 86th Fighter Bomber Group on 17 March 1944. At that time the outfit was based at Pomigliano, an airfield with a hard surface runway about five miles southeast of Naples.

The A-20 was a marvelous airplane. I enjoyed very much the tour with the 47th Light Bomb Group. But there is no flying machine or flying activity that matches combat flying in a fighter-bomber. The A-36 was a great dive-bomber and strafing machine.

Actually, the A-36 was an early version of the P-51 with dive brakes in the wings. Like all of the P-51s, the A-36 had a 37 ft wingspan. Length was 32 ft 3 in. With a height of 12 ft 2 in, the A-36 was eighteen inches lower than later P-51 derivatives. Empty weight was 6,610 lbs. This was 515 lbs less than the P-51D empty weight. The power-plant was an Allison, V-1710-87, 1,325hp, twelve-cylinder, liquid cooled engine.

The A-36 had a good roll rate, good control harmony, and it trimmed well in a dive. Like all liquid-cooled engine aircraft, it was vulnerable to radiator or coolant line hits. Losses in any fighter-bomber outfit are high. The outfits equipped with liquid cooled engine aircraft are even more so.

With the dive brakes extended and the throttle in the wide-open position while in a vertical dive, the aircraft would stop accelerating when the airspeed reached about 375 mph. With the aircraft in this steady-state dive condition, the pilot could concentrate on compensating for wind and eliminating alignment errors. One 500 lb bomb was carried underneath each wing. The A-36 had six 50-caliber machine guns, two in each wing and two fuselage guns. The fuselage guns were synchronized to fire through the propeller. These guns could be hand charged in the cockpit. So, there were no rounds in the chambers of the fuselage guns at takeoff. This allowed the pilot to expend all of the wing gun ammunition, and still have something to use in the event an enemy fighter was encountered after all the wing gun ammunition was gone. The practice also allowed the pilot to attack more ground targets.

There were a few unique missions, but generally missions were one of three types:

- Dive-bombing designated targets

- Low level strafing (troop concentrations, ammunition dumps, airfields, railway marshaling yards, vehicles, etc.)

- Combined dive-bombing and strafing missions

Dive-bombing runs were usually initiated from altitudes between 6000 ft and 8000 ft. The A-36s Allison engines experienced substantial reduction in maximum power available at altitudes much above 10,000 ft. In addition, targets were easier to identify at lower altitudes. The reason for the 6000 ft altitude lower figure was that the twenty millimeter and forty millimeter flak began to be effective at altitudes below 6000 ft.

The German 88s were much less of a threat for fighter-bombers, if you were alert. The muzzle flashes when the guns fired were visible to pilots. So, when any member of the flight saw a flash, he called "flak" over the radio. The flight immediately changed altitude and heading.

Bombs were released at about two thousand feet. This gave sufficient room to pull out of the dive without going much below a thousand feet above the ground. Now and then a pilot would start the pullout too late and incur bomb damage to his airplane.

FAULTY LOGIC

Divine providence intervened on my behalf early in my tour with the 86th. Some mission in the first five that I flew in the A-36 called for us to take out a railroad bridge north of Rome. As "Tail End Charley" in the flight, I was number four in the vertical string of fighter-bombers diving on the bridge. From that position, I had the opportunity to see where the bomb from the first three A-36s hit. All were close....but no cigar. With lots of enthusiasm and zero smarts, I decided that I would hold my bombs, pull out of the dive, and make a low level run perpendicular to the track and below the top of the bridge abutment. Bombs would be released, just before pulling up to clear the bridge. If I had succeeded, I would have blown my aircraft to pieces. Fortunately, the bombs didn't detonate. Most likely, the bombs were released so close to the abutment that they didn't have time to arm before they hit.

TERRY EDMONDS LAST MISSION

Terry went down somewhere south of Florence on his sixth mission. We were in the same four-ship flight. After dive-bombing a bridge, we stayed down on the deck, strafing targets of opportunity. Pickings were very good. There were a fair number of trucks and supply dumps. There was also a fair amount of twenty millimeter and small arms fire. No one saw Terry go down. When we reached the coastline and climbed to join up and head home, he wasn't there. In 1982, acting on a tip from Leon Nowak, an A-20 squadron mate of Terry's and mine, I found his grave in a military cemetery near Florence.

35

CLOSE CALL

Probably the closest I came to buying the farm in an A-36 was on a dive-bombing and strafing attack of the Viterbo Airfields north of Rome. I was leading the second element of a flight. During the dive, I noticed some gun emplacements firing from a small valley among low rolling hills just south of the airfield. Observing that they would be along my flight path during the pullout, I decided to pull out of the dive below the tops of the hills separating the airfield from the gun emplacements and surprise the gun crews with a strafing pass.

It was O.K. to drop up to twenty degrees of flaps at 250 mph in the A-36. My plan was to put down 20 degrees of flaps and come over the tops of the hills with the aircraft nose depressed and all guns firing. I never pulled the trigger. Every gun in the battery must have had their barrels lowered waiting for me to appear. Among the hits was an explosive round that hit the outboard right wing leading edge. The metal from the sizable hole it made peeled back, producing a right roll that almost put me inverted about 20 feet above the ground. Fortunately, I had just enough aileron control to remain upright and get out of the area. I didn't realize that my wingman had followed me. Usually, we were on our own during the dive. We reassembled after clearing the flak area. Looking back, I could see that my wingman's aircraft had made a large fireball when it hit the ground.

MY TERMINALLY WOUNDED A-36

UNIQUE SUPPLY MISSION

The bomb line and the ground battle hung up at Casino for quite a while in the spring of 1944. When we were flying A-20s, Terry and I spent time at a British artillery spotting position on the Allied side of the line at Casino. After we had been there for about an hour, one of the Brits told us that two guys had been killed by snipers a few days earlier, at the point we were standing. We quickly terminated our visit. The Germans owned the high ground, the Monastery on the top of Mount Casino. Several efforts to dislodge them had failed. The Indian Gurkas, a very tough bunch of fighters, made one of the attempts. When they were pinned down halfway up the mountain in need of ammunition, food, and water, our group was called on to attempt a re-supply. We carried bundles of what was needed on our bomb racks and dropped them from heights of less than 100 feet directly over where they were pinned down. We put them where they were needed without losing anyone. However, we never heard whether or not the troops retrieved anything.

HEAD UP AND LOCKED

Sometimes challenging situations were produced by gross inattention to fundamentals of aircraft systems management. On this particular day, the squadron was returning from a mission north of Rome to our steel-plank runway near Caserta. We were buzzing for fun in a loose formation of four ship flights, more or less, in string. For no apparent reason that I could determine, my engine suddenly quit. We were moving fairly fast, so I pulled up to get some altitude and look for a place to land. As I did, the engine started running again. It continued to run for a few minutes. By radio I advised the other squadron pilots that I would appreciate landing priority when we reached the airfield. From my thousand or so feet I could see the airfield a few miles away, when the engine quit again. I knew that I couldn't glide to the runway from where I was; so, I scanned the territory for a closer landing spot. The available territory was not at all promising. As I mentally debated which landing alternative offered the best odds, the engine fired up again. This time, I was able to get to the high key point for a dead stick landing with the engine still running. From there things were business as usual.

By the time I coasted off of the steel mat on to a dirt run-up area, the engine quit and the propeller stopped turning. Sergeant Conrad, my crew chief and an outstanding aircraft mechanic, wasted no time getting to the aircraft. As I verbally gave him a detailed account of the mysterious behavior of OUR airplane, he checked the fuel selector and interrupted my discourse with a wide grin. With just an ever so small hint of scolding in his voice, he said, " If you had switched to the reserve fuel position on the fuel selector when you ran out of fuel in the main tanks, the engine wouldn't have stopped even once." Most likely, if I had been flying any fighter other than the A-36, I would have bellied the airplane somewhere in the boondocks. The reserve fuel system in the A-36 was unique. There was no separate reserve fuel tank. Instead, a standpipe in one of the main tanks limited the amount of fuel that could be used from that tank with the fuel selector in any other position. To burn the reserve fuel, the fuel selector had to be moved to the RESERVE position. What must have happened is that, when the aircraft attitude was changed, some more fuel flowed into the standpipe and the engine started and ran until that amount of fuel was burned. Needless to say, my ego took a solid blow to the midriff.

FATAL LANDING MISHAP

As the ground troops moved up the Italy boot, we changed airfields. As a result, the operational suitability of the airfields varied considerably. Sometimes we operated from a dirt strip or grassy field. At Pomigliano, we had a paved runway. At Caserta, we had a single steel mat. The limited taxiway access to the steel-mat often had airplanes taking off one direction and airplanes landing the opposite direction. Except when a Squadron or the Group was flying a mission, traffic was very light. Mission crews were well briefed on how we would takeoff and land. So, we had no difficulty with taking off and landing in opposite directions. That is, we had no difficulty until the day that my squadron was returning from a mission and landing to the west, while a Beaufighter blocking the turnoff taxiway, waited to takeoff to the east. I was number four in the first flight to land. A dirt apron, large enough for my flight to pull off, paralleled the west end of the steel-mat runway. We four cleared the runway and held on the apron with our engines running.

As the second flight leader, George Simpson, slowed down to clear the runway, there was no place to go. So, he came to a stop. The steel-mat was quite narrow. After touchdown it was best to keep the aircraft rolling in the center. Once the A-36 was in the three-point landing attitude, nothing on the runway ahead of you was visible. Assuming that the number-two flight-leader had turned off on the taxiway, George's wingman let his aircraft roll so that the runway would be clear for second flight element leader. As a result, the propeller of the wingman's aircraft chopped through the empennage, fuselage, and cockpit of George's A-36. George was killed instantly.

TARGETS OF OPPORTUNITY

Now and then, I served as flight leader. Once, I led the Squadron when our Squadron was Group-lead. But about halfway through my tour with the 86th, I started requesting and was given the number four slot in the flight, "Tailend Charley." When the primary objective of the mission had been accomplished and the flight turned for home, I would pull out of the formation and use whatever fuel I had in excess of what I needed to get back to the airfield searching for targets of opportunity. Sometimes there were lots of them. Other times only a few. But every mission was a new experience. On one such foray, I caught a German convoy on a winding mountain road north of Cassino. The road was cut out of the sloping mountainside; so, there was no place for the vehicles to pull off and hide. I had a field day until I finally ran out of ammunition. Most of the German troops scattered and took cover. However, there was one gutsy motorcycle rider who never left his bike. As I made passes at the vehicles in front and behind him, he knelt down and continued firing his automatic weapon at me. I could almost see the color of his eyes each time I came by. Even though his behavior was not exactly friendly, I was sort of pleased that he apparently survived the encounter.

P-40 WARHAWK

There were two A-36 outfits in the Mediterranean Theater, the 27th and the 86th. The A-36 production line stopped early. So, there were no replacement aircraft for the two groups. When the number of available A-36s in the theater became insufficient for two groups, all of the remaining A-36s in the 27th were transferred to the 86th. The 27th was given some war-weary P-40s to use until P-47 replacements came in. As the 86th A-36s continued to diminish in number and the 27th P-47s started arriving, the war-weary P-40s were passed on to the 86th. My squadron was the last squadron in the theater to fly A-36s. Finally, we ran out and started flying the last of the P-40s.

The first P-40 purchased by the Army Air Corp was essentially a P-36 modified for an inline Allison engine installation. The 86th was equipped with hand-me-down P-40s. These had 37 ft 3.5 in wingspans. The length was 31 ft 2 in. Wing area was 236 ft 2 in. Empty weight was 6,480 lbs. Maximum takeoff weight was 8900 lbs. Cruise speed for maximum range was 250 mph. Our P-40s were powered with a Packard/Rolls-Royce V-1650-1 Merlin, 1350 hp engine.

MY P-40
THE STRAW HAT WAS PART OF THE LOCAL DRESS UNIFORM

The P-40 was a fun airplane to fly. It would turn tighter than the A-36. But it was much slower. The A-36 was an aileron airplane. The P-40 was a rudder airplane. Holding the ball in the center when diving the P-40 was a great left leg muscle builder. It could be said of the P-40 pilot that he never ran away from a fight.......everything was faster than a P-40.

Our P-40s Merlin engines had very sensitive cooling systems. We staggered our engine starts. And everyone taxied and took off as quickly as they could. Even with our best efforts, the engines would sometimes overheat before we could get airborne. When the engine overheated, a "pop-off" valve would open to relieve pressure. P-40s rolling down the runway blowing steam was a common sight on a hot day.

Our P-40s carried one 1000 lb bomb between the landing gear. There were three 50-caliber machine guns in each wing. Angle bombing techniques were used instead of diving vertically.

TRADING A P-40 FOR A P-47

A P-40 happening in mid-spring of 1944 almost put me on the ground in enemy territory. The German's started pulling back from Casino, increasing their use of highways during daylight hours. As a result, the 86th group was given several targets of opportunity missions to destroy as much enemy equipment and as many troops as we could. On one of these missions, we carried fuel drop tanks instead of bombs to allow us to strafe as long as we could. I ran out of drop tank fuel about 20 miles northeast of Rome and jettisoned the tanks while I was busy. Several minutes later I checked the fuel remaining in my inboard tanks and found them very close to empty. I never found out what happened to the fuel system to deplete the inboard fuel so rapidly. The home airfield was about a hundred miles away. There was absolutely no chance of getting home or to any airfield south of the bomb line. I remembered that Anzio Beachead had a steel-plank runway. George Simpson and I had once delivered mail there in a "Bamboo Bomber" (C-78).

Anzio was about thirty miles or so south of Rome. I knew there was quite a bit of flak around Anzio Beachhead. On the other hand, the high probability of running out of fuel if I took a circuitous route and approached Anzio from the sea was not at all appealing. I chose to fly directly to the Anzio runway. It turned out that this was indeed the right decision. If there was any flak, I didn't see it. The P-40 engine quit for lack fuel just as I rolled to a stop at the runway turnoff point. There was no power to clear the runway. Ground crews had to push the aircraft off of the steel mat.

I had absolutely no desire to spend time at Anzio. So, I asked one of the ground crewmen who helped push the P-40 off the runway if there was, by chance, any flyable aircraft that needed ferrying south. To my good fortune, there was a 27th Fighter Bomber Group P-47 that had been repaired at Anzio and was ready for pickup. A crew chief showed me how to start the machine. About 45 minutes later the 27th had their P-47 back.

That was my one and only flight in a P-47. Later the 86th was outfitted with "Jugs". By then I had returned to the States.

THE INDIAN MOTORCYCLES

Most of the time I spent with the 86th, I shared a tent with a great guy and fine pilot named John (Ripp) Reichart. Ripp traded two fifths of Scotch to a British sergeant for two Indian motorcycles. This enabled us to travel anywhere in Italy that wasn't occupied by the Germans. We covered a lot of the Allied liberated territory. On one excursion, we found piles of German hand grenades (potato mashers) and several boxes of dynamite. Most of it was used for fun and games. With a small part of our ordnance depot, we went fishing in a lake located on the Pope's summer-house grounds south of Rome. Two or three dynamite sticks tied to a hand grenade blew a lot of water into the air.

I never did become a skilled motorcycle rider. Once, while breezing through an Italian town, on their customary narrow streets, I met four Frenchman in an American Jeep as I rounded a blind corner. Recognizing that a head-on was going to happen, I turned loose of the handlebars and rolled backwards off of the machine. The motorcycle climbed the Jeep hood, went between the four Frenchman who were ejecting themselves from their vehicle, and off the Jeep rear end. I dusted myself off, stared at the Frenchman with my most disdainful stare, mounted the motorcycle, and roared off. Ripp laughed a lot.

8

Back to the USA

SHOT DOWN BY A MOSQUITO

I n late spring of 1944, an Italian mosquito grounded me. Suddenly I was too sick with malaria to care about anything but either getting better or having someone put a 45 slug through my head. I felt like I had been across the North Atlantic in a Liberty Ship during a five-day storm while someone forced me to swallow castor oil every thirty minutes. The Flight Surgeon sent me to a field hospital near Caserta. In a week or so I began to feel much better. I would like to say that I wanted to give my bed to a real patient. There was a major push going on along a line stretching from Casino to the sea. Seriously wounded infantrymen were arriving by the truckloads. The truth is that I didn't care for time in the hospital. I just wanted to get back to the 86th. When I asked to be released, the field hospital management told me that my stay would be more like two more weeks. Since my clothes were hanging adjacent to my bed, I waited till the next day when all of the hospital help was doing serious work, dressed and departed.

The 86th's airstrip was only about fifteen miles down the road. So, it was easy to hitchhike there. The Flight Surgeon was surprised that I had been "cured" so fast. But he didn't press the question about how I recovered so quickly. It just happened that the squadron had a mission the next morning. I was assigned to a slot. I don't recall what we did on the mission. I clearly recall that when I got back on the ground, I was back into the "sick as a horse with malaria" mode. The Flight Surgeon was more than a bit unhappy. In quick time, I was returned

66

to the field hospital. My clothes were locked up. And I stayed a few days longer than the full time required for me to recover.

After the bout with malaria, I was given the option of being reassigned to the States. Since I hadn't been home for three years, I decided that it would be great to make a visit home on my way to either the South Pacific or China. Had I known that the Army had absolutely no intention of letting me go to any other theater, I would never have left the 86th. Too late we get smart!

After I left the Group for Naples to catch a ship to the States, orders commissioning me as a second lieutenant arrived in the Squadron Orderly Room. One day while killing time at the Naples airfield, I ran into some 86th pilots who advised me to get back up to the Group in a hurry and formerly accept the commission. I left Italy without doing so. Later I received a letter from the Squadron Adjutant telling me that the commissioning orders had been revoked. I finished the war as a Flight Officer.

399TH FIGHTER SQUADRON, 369TH FIGHTER GROUP: P-40N

Joining the 399th Fighter Squadron of the 369th Fighter Group eased considerably the disappointment of being stuck with a stateside assignment. The unlimited, unfettered flying opportunities in the 399th were on a par with the opportunities afforded a hungry, sweet-toothed kid all alone in a candy shop.

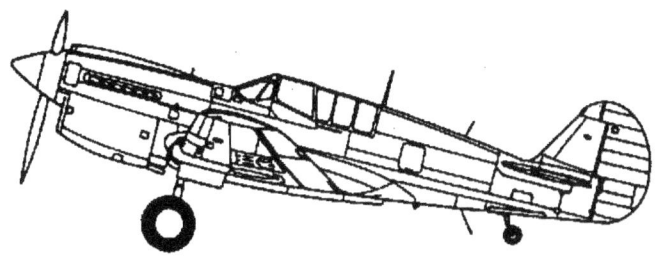

The 369th was equipped with P-40Ns. The P-40N was about two feet longer and 480 lbs lighter than the 86th P-40s. The "N" was powered with an Allison V-1360 hp, supercharged engine. Most of our aircraft were fairly new. When I reported in at Deridder, Louisiana, the 399th had temporarily relocated to a field near Crowley, Louisiana, on a deployment exercise. Since all of the P-40s were at Crowley, I flew the Squadron's instrument training BT-13 to Crowley. The stay there was short. Within a week or so we were back at Deridder.

SERIOUS ERROR IN JUDGMENT FLYING THE BT-13

It was at Deridder in a BT-13 that I made the greatest mistake in all of my flying days! Each squadron of the 369th Fighter Group was equipped with a BT-13 for use as instrument trainers. They were also available for general proficiency flying. In those days, Army Regulations allowed a pilot to take his wife for a flight once each year. Katie and I had been married only a few months. A good friend, Ben Kenyon, had also just married. For whatever reason, we decided one good flying day that we should acquaint our wives with the wild blue yonder by taking them along on a formation flight. We stupidly allowed our egos to overwhelm good judgment and got into a dogfight. The inevitable result was that two first-flighters became very, very flight-sick. Katie has never cared for flying since...... Bless her heart, she has tried from time to time, for my sake, to make it as a flyer...but the bad beginning was too much! The price for some stupidity is high.

NEWLYWEDS '44

DOGFIGHTING

At that time, the 369th Group was made up of about half returnees from combat and half new pilots just out of flight training. A goodly number of the new pilots were recent graduates from West Point. All of them were eager and always ready to fly.

There was a great rivalry between the squadrons. Verbally and in the air every effort was made to give credibility to boasts of being the best fighter pilots on the airfield. Frequently, even before the gear was up and locked in the wheel wells after takeoff, one or more P-40s from the other two squadrons pounced on the disadvantaged no-airspeed, no-altitude addition to the fray. Maneuvering at two-hundred feet above the ground at full throttle, with the airspeed just a few knots above stall, really refined one's low speed maneuvering skills.

Kinder, Louisiana, a little town south of Alexander, was a favorite jousting place for engagements with P-47s based at Baton Rouge. The "jugs" were considerably faster than we were. They were also dead meat if they tried to maneuver with us, which they almost always did.

399TH FIGHTER SQUADRON PILOTS

MORE A-36 FLYING

A few A-36s were parked at an out of the way spot on the Deridder Airfield. I never knew how they got there. The time came when an order for them to be flown to another airfield came down the command chain. Since I was the pilot in the group who had the most time in A-36s, I had the good fortune of test hoping most of them to make sure that they were still flyable. There were a few surprises. As I rotated for climb after one takeoff, a lap full of water poured from behind the instrument panel. With some allowances for out of tolerance conditions, they all flew well enough for one ferry trip.

WIRE CUTTING

A favorite training activity was low level flying, individually and in spread flights of four aircraft. The purpose of the training was to teach the uninitiated how to fly low while visually acquiring targets, without flying into the ground. It was a very important training phase. We lost a few replacement pilots in the 86th Fighter Bomber Group in Italy because they simply flew into the ground. To cure the problem, 86th replacement pilots were required to complete a low-level training program before flying combat.

There were several "rules of thumb" which improved one's chances of not messing up. One was to pull up above telephone pole height when about to cross a road of any kind, whether poles or wires were sighted or not. After many hours of low-level experience, the practice became an automatic reaction. Low-level training of the sort which most of us practiced was not officially sanctioned in the 369th. However, the leaders generally knew what was happening. In fact, it was not uncommon to find many of them participating from time-to-time.

One happy day, while leading a four-ship flight southeast of Deridder, Louisiana on an unscheduled low level exercise, the automatic reaction created rather than avoided a problem. We were all flying line abreast in a spread formation (every pilot flying his aircraft... only occasionally checking the leader's relative location). A tree-lined country road appeared directly ahead. With no lateral deviation, my aircraft's flight path would pass directly through a gap in the tree line. So, the obvious thing to do was to go between the trees. Unfortunately, the automatic reaction to pull up didn't set in until the aircraft was just about to pass through the gap. Had I not reacted, the P-40 would have passed under the wires, unscathed. As it turned out, the propeller spinner hit the center of the wires. The prop cut the wires. And the wings collected the pieces. None of the other pilots saw what happened, as I pulled up out of the formation.

As I watched the three other P-40s continue straight ahead, my thought process reengaged, particularly the memory section of the gray matter. I clearly recalled the official communiqué from the Group Commander stating that the next pilot caught buzzing would be court marshaled. One of the "jocks" in another squadron had knocked a wing tip off of a BT-13 impressing some audience only a few weeks earlier. What to do????? To buy time, I announced over the radio that I was encountering a problem and that I would be returning to Deridder. Refusing an offer for assistance, I told the three to return to Deridder; and I would see them there.

Checking my geographical location, I realized that I was only about five minutes from the empty airfield at Crowley, Louisiana. So, the logical first step was to land at Crowley, pull the wires off of the wings, and assess the damage. If the damage was too severe, I could always fly the aircraft over the swamps south of Crowley and bail out. Fortunately, the damage was such that I never had to find out if I would really have taken the second alternative.

The only one in the Group, other than myself, who ever knew what happened was my crew chief. When I taxied in, he looked at the prop spinner, and grinned as he said, "Looks like you cut a few wires. Don't sweat it, a little brillo and some paint will have it looking like new." The other flight members still think that I fouled the plugs by inadvertently running on one magneto and causing the engine to run rough.

THE GROUP COMMANDER'S A-20

I was pleasantly surprised to discover that the Group Commander, Major Brewer, had his own A-20. In those days, things were different. No one, including the news media, cared that Major Brewer had an A-20 for his personal transportation to and from anywhere he chose to go. And no one cared that I functioned as a ferry pilot for him when he wanted to go somewhere and have me bring the A-20 back to the home airfield for the maintenance crews to work on while he was away.

Major Brewer was not particular about the routes I took returning to the home airfield or bringing the A-20 back to him when he was ready to return. As a result, I had some great times visiting airfields of my choice. On one flight I was able to fly over my hometown, Batesville, Arkansas, and blow the dust off of the roof of the McElmurry Grocery Store located in West Batesville.

Only once did I spend a night away from Deridder on an A-20 trip out and back. The overnight stay was not by choice. Major Brewer's A-20 had only one navigation radio, a low frequency range receiver. Deridder had no low frequency range. I rode in the A-20 nose, with Major Brewer piloting, to Maxwell Field,

Alabama. We landed there just about dark. A cold front was forecast to be in place at Deridder by the time I arrived on the return leg. I recall the Base Operations Officer at Maxwell Field Operations asking me about my instrument rating, and the lengthy discussion we had when I told him I had one, but it was out of date. He was not at all pleased with my proposal to file VFR. I finally convinced him that my recent experience in Italy, where no one concerned themselves with instrument cards, should be acceptable evidence that I was qualified to make the flight. About an hour later, I found myself at about five hundred feet above some little Louisiana town, just below the clouds, trying to stay VFR in a driving rain with turbulence and lightning all around. I acquired an instant appreciation for the Maxwell Base Operations Officer's viewpoint. Conditions in the Deridder direction looked even worse. So, I decided to fly due south and try to pick up the light line into Lake Charles. Again, I was blessed with good fortune. I did, indeed, pick up and fly to an airway beacon.

However, a new problem developed. I turned to the airway heading and began looking for the next beacon. When I finally acquired it visually, a turn to the right of more than forty-five degrees was necessary to fly to it. This entered a question into my mind as to whether or not I was on the light line to Lake Charles. When the sequence repeated itself again, getting to the next beacon, I wasn't sure what to believe. A crosswind was my first thought; but I couldn't imagine a crosswind that strong. I finally saw the rotating beacon at Lake Charles and the crab necessary to get there confirmed that the wind out of the north was what it appeared to be, unusually strong. I was one happy pilot when the gear touched the Lake Charles runway. I slept on a couch in Lake Charles Operations.

After leaving the 369th, I never had another opportunity to fly the A-20.

TEMPORARY FERRY DUTY: P-51Ds

I was one of about five 369th Group pilots who were sent to Long Beach, California, in January 1945 for a month to ferry P-51Ds from the West Coast to Newark, New Jersey. At Newark, the wings were removed. The airplanes were then cocooned to protect them from salt spray. Cargo ships then hauled them to England.

About three days were required to fly a P-51 from Long Beach to Newark and return to Long Beach for another one. Riding a commercial DC-3 from Newark to Long Beach consumed one of the days. Weather and refueling stops used up the other two. Katie spent January with her parents in Little Rock. Little Rock became a mandatory refueling stop. Ferry pilots were not allowed to fly at night. Unless there was enough of the day remaining when I reached Adams Field to make it all of the way to Newark before sundown, Little Rock was also the overnight stop location.

If enough daylight was left when I arrived at Adams Field to make one more leg, but not enough to get to Newark, I made certain the overnight stop would still be at Adams Field. This was done by lengthening the ground time at Little Rock by running on only one magneto for half of the inbound leg to Adams Field. The fouled plugs ensured that a magneto check would exceed tolerances. Plugs would have to be cleaned before proceeding. After two such events, the line chief at Adams Field cornered me and made a proposal I couldn't refuse. The options were to simply run on two magnetos and let the line chief make an entry in the log that the plugs had been cleaned.......or, experience a very low priority in all of the service provided at Adams Field. Easy choice!

369TH GROUP DUTY AT STUTTGART ARMY AIR FIELD

While I was TDY ferrying P-51Ds, the 369th Fighter Group moved to Stuttgart Army Air Field in the Arkansas flat lands. By this time, the Group pilot population was about seventy percent combat returnees and thirty percent recent flying school graduates. We rejoined the 369th Fighter Group in February 1945.

Flying at Deridder had been great. It was unbelievably great at Stuttgart:

- Most of the returnee pilots in the Group had no great desire to fly every day. This meant more airplanes for the eager beavers.

- Stuttgart Army Airfield was less than 15 minutes flying time from my hometown, Batesville, Arkansas.

- Batesville had a small airfield that easily accommodated an L-5.

- In addition to many relatively new P-40Ns, there were L-5s, P-51Bs, BT-13s, and one B-25 available for flying.

The candy factory had increased in size, it was located in my backyard, and its doors were unlocked eighteen hours of each day. It can't get any better than that!

TYPICAL FLYING DAY AT STUTTGART

A typical day at Stuttgart began with a dawn patrol of the air space over and around Batesville. The order of the day didn't vary much. To let the civilians know that their defenders were alert and on duty, a wake up acrobatic routine was performed just outside the city limits. Imminent departure from the area was communicated with a shingle-lifting buzz job of the McElmurry residence in West Batesville. Theil was always in the back yard waving a dishtowel or apron at the time of departure. A low-level reconnaissance of White River from Batesville to Lock and Dam Number Three (about twenty miles north of Batesville) was usually made before returning to a loitering position over Stuttgart Army Air Field. The loitering altitude was about 2000 ft. This provided a good speed and altitude advantage for engaging late sleepers from other squadrons just getting airborne.

Dog fighting took up most of the remaining daylight hours. If there was an overcast with a ceiling of at least 1200 ft, we climbed up through it and had at each other on top. The standard let down was to Split "S" into the clouds and make a straight and level recovery somewhere between 800 ft and 1000 ft. Except for the fact that we were in the clouds without an instrument clearance, this disengagement technique was not particularly hazardous. The territory for fifty miles in any direction from Stuttgart is flat as a board. There were no towers above five hundred feet or tall buildings in the local area.

UNPLANNED FORMATION WITH A B-24

On a day when the overcast was about 4000 ft thick, a Split "S" disengagement resulted in an unbelievable experience. The ceiling lowered during the time that we were dog fighting on top. When I completed the recovery to straight and level at the bottom, I was still in the clouds. No problem. I would simply make a 500 ft/min descent until I broke out. I just happened to take my eyes off of the instruments for a second, just long enough to discover that I was flying formation with a B-24. There was no more than 50 ft between my left wing tip and the B-24's right wing tip. I stared with amazement at the B-24 copilot and he returned the stare with a similar expression of disbelief. The probability that we would be at the same point in space at that moment, flying parallel flight paths at the same altitude and airspeed aren't too much better than the probability that you will win this week's lottery. Divine intervention took care of a bunch of us that day.

P-40 KILL.......THE ENEMY IS US!

I managed only one aircraft kill in World War II. I shot down the P-40 I was flying.

One of the things the 369th Group did was support the training of infantry recruits and draftees. A very important training activity conducted at Camp Shelby, an Army training base in Mississippi, was to condition trainees with field exercises using live ammunition, real flame throwers, live grenades, and such. The exercises were conducted on a field about half a mile wide and about three-quarters of a mile long, surrounded by trees. From a departure line on one end of the field, a platoon or company of infantrymen would crawl underneath barbed wire for the length of the field while 50 caliber machine guns fired life ammunition a few feet above the tops of the barbed wire. At just the right time to provide maximum shock effect, mines with lots of bang but no shrapnel, were remotely detonated by instructors. There were concrete pillboxes about midway from start-to-finish on which the troops used their flame-throwers. I was one of three P-40 jocks sent to provide a week's worth of support to a series of these training exercises. We flew out of the City Airport at Gulfport, Mississippi. Our role was to provide friendly ground support when the troops were about a fourth of the way from the departure line. One P-40 carried air-to-ground rockets fired from tubes mounted underneath the wings. The other two P-40s carried practice bombs (100 lb blueboys). All three aircraft carried a full load of fifty-caliber machine gun ammo. We alternated roles.

The routine began when we arrived over the training field at about three thousand feet. We remained in string throughout our time over the target. The two P-40s carrying bombs led the first pass, dropping their bombs on the pillboxes. The P-40 with the rockets, flying Tail-end Charley, spaced himself such that his rockets were fired soon after the number two aircraft had initiated his pull-up. After our external ordnance had been released, we stayed in a circular pattern, making strafing runs about 50 feet above the barbed wire, until all of our ammunition had been expended. The empty shell casings from our guns would fall down on the troops, adding even more realism to the exercise.

About the third day we were there, my turn to fire the rockets came up. Having given some thought to just how this should be done, I planned to put the rockets in the small entrances of the pillboxes. The passes would be slightly below the top of the pillbox. Everything went as planned except that at least one of the rockets made its way up through the P-40 engine. It was readily apparent that flight was about to come to an end. Further, the ending was going to be in the trees. I quickly rolled the canopy back, zoomed the aircraft, unfastened my seat belt, and tried to stand up. The wind instantly shoved me back down into the seat as the P-40 nose dropped below the horizon. With an altitude of less than 800 ft, instant departure from the cockpit was absolutely essential to survival. Grabbing the left side of the cockpit, as I turned sideways, I somersaulted on to the wing and slid off of the trailing edge. Fortunately, I cleared the horizontal tail. I clearly recall looking down at the D-ring, grabbing it with my right hand, and pulling it to full arms length. Then, looking at the D-ring, I silently said to myself, "You're dead. It didn't open in time." The thought had barely made it through my mind when there was a popping sound like a balloon bursting and my legs were snapped to a fully extended position. What a wonderful feeling!

The parachute and I made one full swing and half of another before we hit the ground in the middle of a blackberry patch. By the time I extricated myself and the parachute from the blackberry patch, an Army Jeep with a second lieutenant driving was coming down a nearby dirt road. He was quite surprised to see me. Their line-of-sight, visual contact with the airplane had been blocked by the trees before I managed to somersault out of the cockpit.

I had made plans to fly to Stuttgart, Arkansas that evening to spend the night with Katie, and did make the trip in one of the two remaining P-40s. However, it required three tough three-minute rounds with the Flight Surgeon before I convinced him that there was absolutely no reason that I should have a flight physical before flying again.

B-25 CROSS-COUNTRY TO CHICAGO

I never learned what happened to Major Brewer's A-20 when the 369th Group moved from Derrider to Stuttgart or how the Group acquired a B-25 after the move. But the Group did have a B-25. There came a time when a non-rated officer from the Chicago area convinced whoever was in charge of the B-25 that he should have B-25 transportation to and from Chicago.

NORTH AMERICAN B-25
MEDIUM BOMBER CARRYING A CREW OF FIVE

B-25 pilots were in very short supply in the Group. In fact, I never met a pilot at Stuttgart who had flown a B-25. In desperation, the non-rated officer, acting on a report that McElmurry was an A-20 pilot, looked me up and asked if I could and would make the trip. Of course, I would. I had never been checked out in a B-25. But I had been given an instrument card after one twenty-minute instrument training flight in a B-25 at Will Rogers Field in 1943. I don't recall any difficulty in convincing the owner of the B-25 that I was qualified to fly it. In fact, I don't believe anyone asked.

A co-pilot was required. So, I asked Ike Gugler, a 399th squadron buddy and an excellent pilot, if he was interested. He was. This was Ike's first time in a B-25. However, like most fighter pilots, Ike's attitude was "It's an airplane, isn't it?"

We departed Stuttgart with three passengers, about mid-morning, for Bowen Field, Kentucky. As we approached the field for a landing, we observed a squall line approaching the field from the north. The line of thunderstorms extended as far as we could see to the west and the same distance to the east. It was clear that if we intended to go to Chicago from Bowen Field, we would have to get on the north side of the thunderstorms some way. In retrospect, the logical thing would have been to land, refuel, and wait a few hours until the squall line had passed Bowen Field. That was not the choice we made. Instead, we refueled as quickly as we could and took off. During the stop we picked up three more passengers who were hitching rides to Chicago.

As we climbed in the clear airspace south of Bowen Field, the squall line began to pass over the airfield. Any thought of returning to Bowen Field was now definitely out of the picture. Hoping for a gap between the thunderstorms, we climbed to 15,000 ft flying parallel to the squall line. Not only were there no gaps, the tops of the thunderstorms were obviously well above any altitude that we could hope to reach with a B-25. Since neither Ike nor I could come up with any better idea, we agreed to an attempt to go under the thunderstorms. So, we let down to an altitude of about eight hundred feet and took up a course which was ninety degrees to the boiling roll cloud.

Only the structural over-design by the North American design engineers saved the B-25 from instant destruction as we passed under the roll cloud. Maintaining a heading was a distant second priority to keeping the aircraft semi-upright. The aircraft gyrations were in all directions. Most of the abrupt aircraft center of gravity motion was up and down. By this time, it was very clear that we had selected an option other than the best one available to us. In an attempt to make both Ike and myself feel better about our situation, I assured him that when we hit the rain, things would get smoother. Very quickly, this distortion of fact was exposed as pure hogwash. Turbulence did not diminish in the least as we entered the torrential downpour. Except for frequent bolts of lightning, our daylight VFR flight instantly changed to mid-night black instrument conditions. We had no idea where we were geographically. So, we climbed to about 1500 ft and held a compass course in the general direction of Chicago. In about 15 minutes, the violent motion of the flying machine began to subside. In about another 20 minutes, we broke into the clear. The remainder of the flight to Chicago was uneventful.

The flight home began late in the afternoon of the next day. The Missouri-Arkansas-Tennessee part of the country was forecast to have generally low ceilings with drizzle throughout the night. As it turned out, the weather prophets hit it on the head for a change. Since Stuttgart had no instrument approach, the plan was to fly to Little Rock then, take up the compass heading to Stuttgart. We would then slowly let down until we broke out of the clouds. Stuttgart was the only town of any size in the general area. So, we should be able to visually acquire the town lights and the airfield rotating-beacon. When we arrived over the Little Rock low frequency range, the Little Rock Radio communicator advised that the ceiling at Little Rock was above minimums. There was no one ahead of us, if we wished to make an approach into Adams Field. Ike was in favor of taking Little Rock Radio up on their offer. However, after a brief review of the advantages of sleeping in our own beds, we agreed to continue with the earlier plan.

We may have used a faulty compass heading for the letdown to Stuttgart. Or the wind may have blown us off course. For whatever reason, we broke out over

Memphis, Tennessee. Ike favored a return to Little Rock and a good night's sleep there. The time was close to midnight. He agreed to one more try for Stuttgart on the way to Little Rock.

As we passed through 500 ft. on the altimeter letting down in the direction of Stuttgart, it occurred to me that I had not reset the altimeter since we left Chicago. Our actual altitude could be three hundred feet above the altimeter reading... or three hundred feet below the altimeter reading. It was obvious that Ike was somewhat disturbed that we were still in the clouds at an altitude of 500 ft. So was I. To lessen Ike's anxiety, I assured him that the clouds never went all of the way to the ground around Stuttgart when Little Rock was at or above minimums. There was no response to my statement. The altimeter needle passed through zero with nothing being said. Finally, with a minus one hundred feet reading on the altimeter, the lights of Stuttgart came into view. From a traffic-pattern about three hundred above the ground, we made our way through the drizzle to a landing at Stuttgart Army Airfield.

I didn't fly the B-25 again until I attended Test Pilot School in 1956.

MY FIRST FLYING STUDENT: BT-13

It was in this "fly as much as you want to" environment that I became acquainted with the 399th Squadron's Line Chief, Sergeant Lilly, a very capable non-com-missioned officer. Sergeant Lilly wanted to fly just about as much as I had as a teenager in Batesville, Arkansas. With flying all around him, he was totally frustrated. Noting that I seemed to be flying from dawn to ten o'clock at night just about every day, he asked if it would be possible for me to take him up in the BT-13 rear seat once in a while. I told him that we could do that occasionally when he had a spare hour or two. Overnight, Sergeant Lilly had a spare hour or two any time of the day that I could find the time to fly the BT-13. He always had the BT-13 fueled and in a flight-ready state.

I had never done any flight instructing. But it was easy to become a flight instructor with Sergeant Lilly as the student. He had an aptitude for flying; and he was a quick learner. In a very short time, I moved him from the rear seat to the front seat. I was ready to solo him. But a much wiser friend, who had been observing this unauthorized activity, strongly advised me not to. Sgt. Lilly had to settle for knowing that, without question, he had become a capable pilot. He would have to wait until some later time to make his first solo flight.

CHANGING SEATS IN THE BT-13

Among the recent flying school graduates in the 369th was an excellent pilot named Kelsey. We spent many flying hours dog fighting and having a flying good time in general. Once, back at Deridder, when I was on Kelsey's tail in a dogfight at an altitude of 1000 ft., Kelsey rolled into a Split-S, pulling straight through. I pulled up and rolled over to see him hit. But it wasn't his time! With wing tip streamers all the way, Kelsey pulled the P-40 out below the treetops. He flew straight and level directly to the airfield and landed.

I don't recall where the idea came from. But, one day when things were a bit slow, we decided that changing seats in the BT-13 in flight was something that we should try. We took off with Kelsey in the front seat and me in the back seat. At about 2000 ft, Kelsey opened the front cockpit canopy. I opened the rear cockpit canopy. While Kelsey held the airspeed stabilized at about 10 knots above stall speed, I unfastened my safety belt and climbed out on the wing. The plan was to let Kelsey unfasten his belt and rise up enough for me to reach under him and take the stick. It turned out that we couldn't manage this with me on the wing. So, after some verbal real-time re-planning at the top of our lungs, I climbed up on top of the fuselage and straddled the canopy behind Kelsey. The revised plan was for me to lean forward and reach between Kelsey's legs, taking the stick as he stood up as much as he could. After we had made a couple of unsuccessful tries, I suggested that I go back to the rear cockpit. However, Kelsey was not ready to fold. So, with greater vigor he pressed forward as he lifted himself upward partially out of the cockpit. It was enough. I managed to grasp the stick and take the airplane.

Unfortunately, as Kelsey moved laterally out of the front cockpit, his foot hit the throttle and moved it to idle. Since we were very close to stall speed, the aircraft stalled. As it turned out, Kelsey's hold on the airplane and my hold on the airplane were solid enough to keep us attached while I pushed the stick forward to make a stall recovery. I scrambled into the front cockpit. Kelsey climbed into the rear cockpit.

Kelsey was killed a few years ago when he fell into a grain elevator while grain was being loaded into it. It was his time.

MUSTANG ADDITION TO THE 369TH INVENTORY

In the summer of 1945, the 369th group received some new P-51Ds. This changed the character of the dog-fighting. The P-40s could easily turn inside the Mustangs. But the P-51s had the choice of starting or breaking off the engagement at will. The P-40N's acceleration and climb performance were definitely inferior to that of the P-51D.

Except for height (the P-51D height was 13 ft 8 in.), The dimensions of the P-51D were the same as the A-36 dimensions. The P-51D empty weight was 7,125 lbs. This was 515 lbs more than the A-36 empty weight. The P-51D's Packard Merlin v-1650-7 engine delivered 1,490 hp. The P-51D was about 75 mph faster than the A-36.

North American P-51D "Mustang"
USAF Museum

Going suddenly from idle to full power at low speeds in a tight turn could produce an instant torque roll. At Stuttgart, one of our pilots gave a demonstration of this maneuver while trying to avoid overshooting the final turn to touchdown. Contact with the ground tore up the airplane. Miraculously, the pilot survived.

About three years would become history before I again flew the P-51D with the Texas Air National Guard.

9

You're in the Air Force Now

END OF THE WAR SEPARATION FROM SERVICE

WWII ended in August of 1945. That event ended almost five years of active military service, which began with the departure of Company L, 153rd Infantry from Batesville, Arkansas, on 23 December 1940. What an unforgettable adventure and memorable life sharing with great people those years were!

I was given a reserve second lieutenant commission as part of the separation process. Katie and I relocated to a single bedroom in the Matthew's Boarding House located in Tuscaloosa, Alabama. There, I enrolled as a Mechanical Engineering freshman in the University of Alabama on 26 August 1945. We had no car and very little income. Absolutely zero funds were available to do any flying.

Fortunately, there was an Army Air Corp Reserve Unit at the Birmingham Municipal Airport. Birmingham was approximately an hour's drive north of Tuscaloosa, along a well-traveled road. About one weekend each month, after the last Friday class, I would hitchhike to Birmingham and spend Saturday and part of Sunday flying T-6s. When Friday afternoon turned out to be a good time for catching rides, I would occasionally be able to get in a flight Friday evening. A couch in Operations was available as a place to sleep Friday night and Saturday night. So, I was able to start flying early and quit flying late. Sunday afternoon, I hitch-hiked back to Tuscaloosa.

TEXAS AIR NATIONAL GUARD DUTY

After earning a degree in Mechanical Engineering, I accepted a job offer from Standard Oil Company of Indiana. The work location was Hastings Oil Field near Pearland, Texas. Katie, and I found a suitable duplex on the outskirts of Pasadena, Texas. Terry was two years old. Steve had just arrived on the scene.

Ellington Field, home of the Texas Air National Guard 111th Fighter Squadron, was located between Pasadena and Hastings Field. As it turned out, they had an opening for a second lieutenant. So, in a matter of a few weeks, I was back in P-51Ds. In fact, I had my very own airplane with my name on the cockpit edge.

I thoroughly enjoyed flying with 111th Fighter Squadron. The route from Hastings Field to our duplex in Pasadena went by Ellington Field. So, stopping for an hour on the way home to practice acrobatics and landings in the Mustang became part of the daily routine. Weekend cross-countries could be taken just about any time a pilot desired. Occasionally, the entire squadron would take a cross-country or fly to Matagorda Island for a fishing day. Very few part-time flying outfits ever had it so good!

A few months later I would join the 27th Fighter Wing. They were not a P-51D outfit. But they did have one that was available now and then for fun flights.

BACK TO THE AIR FORCE

About a month before graduation from the University of Alabama, an Air Force Recruiting team appeared on the campus looking for engineering graduates interested in a Regular Air Force Commission. I filled out their one-page application, signed it, and forgot it. Three months later, I received a wire offering me a Regular Air Force Commission as a second lieutenant. I responded with a conditional letter of acceptance, requiring assurance that my first assignment would be to a fighter wing as a pilot. The Air Force reply was, "There are no guarantees; but there is a high probability that the first assignment will be to a fighter wing." In January 1949, Katie, Terry, Steve, and I left for Kearney Air Force Base, Nebraska to join the 27th Fighter Wing.

Terry was three years old. Steve was six months old. The roads north of St. Louis, Missouri, were a solid sheet of ice and snow from St. Louis to Kearney. If there was another car moving on the highway, I don't recall seeing it. We did see several eighteen wheelers in ditches and on their side. We slid down hills as I tried to accelerate enough to coast over the next one. Fortunately, there was no one in front of us. I recall recovering from a 360 deg spin without losing the forward momentum we needed to get over the next rise. GOD took us to Kearney that day, not my driving skills.

On arrival at Kearney, we discovered there was absolutely no place to live. We finally located a basement for rent (Not a basement apartment... a basement) in Grand Island, Nebraska. A coal bin took up a fourth of the basement. The fourth of the basement next to the coal bin was open with a water drain in the middle of the open floor space. Our shower was a water hose hung from the basement ceiling over the water drain. We occupied the other half of the basement. Our refrigerator was the snow outside a 12"x 30" basement window. Katie and I had a double bed in one corner of the basement. Steve slept in a bassinet on top of a dresser. Terry had a bed in the "kitchen/living space." Describing our quarters as "minimal living accommodations " is a gross overstatement of the quality.

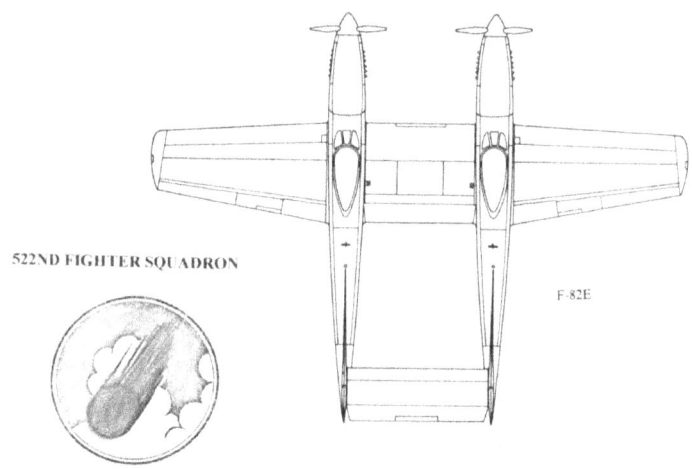

522ND FIGHTER SQUADRON

F-82E

Those of us who resided in Grand Island traveled the forty miles to and from Kearney AFB each day in a tarpaulin covered truck. Most days, the outside temperature was well below zero. But Katie had the worst of it. Shortly after I reported in, our Group Commander, Colonel Cy Wilson, decided to take the group on a round robin to Puerto Rico, Panama, Jamaica, and back to Kearney. For just about all of February 1949 we flew around the Caribbean and basked in the tropic sun, while Katie and other folks in Nebraska "survived" in the sub-zero weather.

The 27th Fighter Wing was equipped with F-82E day fighters. The F-82E looked like Siamese P-51s. Actually, each of the twin fuselages was much larger than a P-51 fuselage. The "E" model was powered with two Allison V-1710-143/145, 1600 hp, supercharged, liquid cooled engines. The F-82 had a 51 ft 3 in wingspan. The wing area was 408 ft^2. The length was 42 ft 2 in. Empty weight was 14,914 lbs. Maximum takeoff weight was 24,864 lbs. Maximum range was 2,708 miles. Service Ceiling was 40,000 ft.

Fortunately, the tour of duty at Kearney was only a couple of months long. On 2 March 1949, we flew the F-82s to Bergstrom AFB, Texas, and left Kearney to the snow and the coyotes. The F-82, like the B-36, was a parting shot by the designers at producing propeller driven combat aircraft. Prototype jet fighters and bombers were already flying. It was clear that jet aircraft would be the combat aircraft of the future. Still, the F-82 was a great performer. At airspeeds well above the long-range cruise speeds of P-51s, we could fly from Bergstrom AFB to Puerto Rico without refueling. The aircraft was an excellent gun platform. There was absolutely no yaw in a dive.

The F-82 would only spin if you forced it into one and held it there with the flight controls. If you relaxed your pressure on the flight controls even a little, the aircraft would immediately stop spinning. The F-82 was a good acrobatic aircraft. Slow rolling the fuselage you were in gave the other pilot a ride around a circle centered on your fuselage centerline. I never snap rolled the F-82. It was probably stressed to handle a snap roll; but it didn't look like it was.

There was one problem with the F-82 that required a lot of attention. The Strategic Air Command F-82s of the 27th Fighter Wing were fitted with two 310-gallon jetisonable drop-tanks, one under each wing. When the tank drop circuitry was armed, all that was needed to jettison the tanks was to brush the bomb release button, located on the top of the control stick, with your hand. If you did, the tanks immediately departed. The tank jettison arming switches were located near the cockpit floor in the forward left corner of the cockpit. The obvious reason for locating them there was to minimize the possibility that the pilot would inadvertently put them in the wrong position. Unfortunately, the cockpit layout expert placed two other similar switches, which were frequently used for normal operations, very near them. Many drop tanks bit the dust as pilots made their pre-start checks and engines were started. In fact, so many tanks were inadvertently dropped on the parking ramp that the Group Commander decreed that any pilot who dropped tanks unintentionally would pay for them.

To minimize these expensive and embarrassing happenings, an inventive maintenance man fabricated a slotted, spring-loaded cover that held the drop-tank arming switch in the "OFF" position. Supposedly, as long as a visual check confirmed that the arming switch was in the slot, the tank jettison circuit was not armed. To arm the tank jettison circuit, the pilot had to intentionally lift the cover, and place the switch in the "ARMED" position. As things worked out, it fell my lot to prove that the cover design was faulty.

Supporting the family on a second lieutenant's pay left no surplus cash to pay for drop tanks. So, during all engine pre-start checks in the F-82, I made doubly sure that tank jettison switches were in the cover slots. The cover check was very positively made during the pre-start checks prior to departing Ellington Field for Bergstrom in the Fall of 1949. Everything went normally during taxi, takeoff, and through gear retraction. Just as I started a climbing turn, there was the unmistakable lateral wing rock characteristic of departing fuel tanks. I immediately half rolled the aircraft to check. Sure enough, the two 310-gallon tanks were tumbling towards the ground. Visions of burning houses, cars, and people flashed through my mind. All I could do was say a quick prayer and hope. If the event had occurred in the 1990s, there would almost certainly have been serious damage or loss of life. Happily, in 1949 the territory immediately north of Ellington Field had many sizable open fields. The tanks landed in the middle of one of these. Quick prayers for assistance also work!

A post-flight check revealed that the protective covers had been fabricated from soft aluminum. A force of some kind had deformed the covers, causing them to hold the switches in the "ARMED" position instead of preventing the switches from being placed there inadvertently.

BACK TO THE BOOKS!

In early Fall of calendar year 1949, the Air Force Institute of Technology announced an opportunity for officers with engineering degrees to earn an Engineering Master's Degree at the University of Michigan. The two-year program focused on engineering related to guided missiles. Ordinarily, nothing but a better flying job than the one I had would get even a look at what was being offered. However, 1949 had been an unusual year. For openers, the Secretary of the Air Force (I believe his name was Johnson.) decided that budget cutting was in order. He decreed that one of the cuts would be in flying time. I suppose the logic was less maintenance and less fuel cost less money. Whatever the rationale was, pilots in fighter squadrons were to be limited to 200 hours per year. Ten of the Group Commander/s cross-country marathons would eat that up in six months.

My flying was diminished even further by an assignment to move a large metal hangar from an abandoned Army Air Base at Childress, Texas, to Bergstrom Air Force Base. To accomplish this feat, I was given two cooks, a medic, a crane operator, a truck driver, and six other airmen who had never done even one day's iron work (neither had I.) We were equipped with one set of car hand tools, a truck with a lowboy trailer, a Quickway crane, and several hundred lengths of glider tow-rope. I had spotted the tow-rope lying on the ramp at Bergstrom just before we departed for Childress. The fact that we managed to get the hangar disassembled and moved to Bergstrom in about four months without killing somebody was nothing short of a miracle.

We had many interesting, sometimes funny, and often very risky experiences disassembling and moving the hanger. The last step in the process was taking the center section pieces of the bowstring trusses down the Texas highways on our lowboy trailer. The center section pieces were too large to travel by rail. They also took up both lanes of the two-lane highway. I led the convoy down the two-lane highway from Childress, Texas to Abilene, Texas in my green Plymouth. Driving squarely down the center of the highway, I would chase all oncoming traffic on to the shoulder. This worked reasonably well until we encountered a Texas Highway Trooper. From that point he led us to the Fairgrounds at Abilene and ordered me to keep everything there until a suitable escort was obtained.

THE DISASSEMBLY & TRANSPORTING TEAM

"BOOMING UP" AND LOWERING
THE TRUSS TO THE FLOOR

The hanger relocation task was followed almost immediately by an assignment to Matagorda Island Airfield as Post Engineer to rehab the utility systems and complete a ground gunnery range. When I arrived at Matagorda Island, I discovered that the water supply system, the sewage disposal system, and the electrical power generation plant were not functioning. To add to the challenge, I was notified that the 100/130 aviation gasoline dump, which had been non-functional since WWII, must be ready in a matter of a few weeks to provide fuel for aircraft arriving to shoot gunnery.

Five merchant seamen were hired off the Corpus Christi docks to help the small contingent of airmen available to get everything done. They weren't much to look at. But they were dedicated and worked like Turks. To my amazement, everything, including the completion of the ground gunnery range, was accomplished on a schedule that supported the 27th Fighter Wing operational requirements. To partially compensate for the fact that I was doing almost no F-82 flying, I was given my own T-6 to do whatever flying I wanted to do, when time was available. There was very little time available.

Peace around the world appeared to have set in for at least a couple of years. So, two years of academics began to look like an acceptable substitute for limited flying with the 27th Fighter Wing and lots of additional duty as Post Engineer at Matagorda Island. So, in the Spring of 1950, Katie, Terry, Steve, and I moved to Ypsilanti, Michigan. I enrolled as a graduate student at the University of Michigan.

We had barely gotten the furniture into our rented house when the North Korean military moved south to take over South Korea. Like me, Hal Fitzpatrick, a fellow student and good friend, had left a fighter squadron to come to Michigan. In short order we scheduled a T-6 at Selfridge Field for a flight to Wright-Patterson AFB and an interview with Colonel Leighton I. Davis. At the time, Colonel Davis was Commander of the Air Force Institute of Technology. Later he would become the Commanding General of the Air Force Development Center at Holloman AFB, New Mexico.

Fitz and I pleaded for a transfer back to our respective fighter groups. We offered to repay all of the costs associated with bringing us to the University of Michigan and whatever money was required to return us to our outfits. Air Force undergraduates at Michigan University were being sent back to operational squadrons in bunches. Still, all of the persuasive arguments we could muster didn't change our graduate student status. We had stuck our feet in a two-year bear trap!

Proficiency flying while we were at the University of Michigan was done at Selfridge Field on weekends and non-class days. Selfridge Field is located north of Detroit, near Mt. Clements, Michigan. We started flying T-6s. Then we changed from T-6s to C-45s. Finally, three of us (Jack Crouch, Hal Fitzpatrick, and myself) talked the 172nd Michigan National Guard Squadron, which was stationed at Selfridge, into letting we three sit their weekend alerts in P-51Ds. We did many practice scrambles each weekend.

THE ZERO-ZERO T-6 LANDING

When flying T-6s, an efficient way to get flying time was to take a round robin cross-country from early Saturday morning to Sunday evening, spending time on the ground only long enough to eat and sleep. I have forgotten what the takeoff airfield was. But, on one of these weekend cross-country exercises I was flying into Little Rock, Arkansas, at midnight on a Saturday. The plan was to sleep at Little Rock until daylight, and then go again. The landing was definitely going to be made at Little Rock, and soon. The fuel remaining was enough for minutes, not fractions of an hour.

As I was letting down to enter the traffic pattern, I requested landing clearance from Adams Field tower. The tower operator advised that the field was below minimums with zero ceiling and zero visibility. Looking straight down, I could see the glow from the runway lights through the clouds. At about 1000 ft, I was clear of clouds. The fuel state dictated one of two options:

1. Ignore the tower operator's weather report and land at Adams Field.

2. Fly to some unlighted spot as far away from Little Rock as the remaining fuel would take me and bail out.

The choice to land took about two seconds to make. The tower operator advised that he couldn't clear me to land. I informed him that his authorization was no longer a requirement. All I needed to know was which runway he would prefer me to use. From 1000 ft altitude to about 300 ft, there was no difficulty staying lined up with the glow from the runway lights. But, suddenly at 300 ft, the visibility did indeed go to zero. No matter, with the existing fuel state, there was going to be a touchdown of some sort.

T-6

Heading and rate of descent became the prime concern for the next few seconds. Holding the aircraft sink rate on the low side of 500 ft/min and the heading deviation to zero, there was nothing to do but wait for contact with the ground. Fortunately, ground contact was made only a few feet laterally from the runway center stripe. Otherwise, there would have been no directional reference. Forward visibility was less than fifty feet. Not knowing how far down the runway touchdown had occurred, I braked to a stop as rapidly as I could, without nosing over. I then advised the tower that I needed a "Follow Me" vehicle to lead me to the ramp. Eventually a vehicle showed up; and we slowly made our way to a parking spot. As we were making our way to the ramp, an American Airlines DC-3 contacted Adams Field Tower about landing. The Tower Operator advised that the ceiling and visibility were zero, but that a T-6 had just landed. The American Airlines crew opted to give it a go. They proceeded to their alternate after one pass.

One set of footprints in the sand that night.

GEAR UP AND FROZEN

Winter flying at Selfridge in T-6s could sometimes get interesting when the weather turned foul. A somewhat challenging weather situation was generated by a combination of instrument weather and icing conditions.

One cold, snowy, winter Sunday at an airfield in the Chicago vicinity, I was preparing for the final leg of a weekend round robin. There were no reports of icing along the route to Selfridge. However, there was a heavy snowstorm in progress there. The Selfridge clouds were solid from about 800 ft to 10,000 ft. This was an acceptable weather state for the T-6. So, into the "muck" I went. En route it was colder than the lower part of an Alaskan well digger's anatomy. Otherwise, the flight to Selfridge was uneventful. Quite a few aircraft were stacked over the airfield when I arrived. So, it took about 45 minutes for my approach time to come. On the down wind leg of the GCA pattern, I was having visions of a hot cup of coffee when the controller told me to configure for landing. Suddenly, things became non-routine when the landing gear warning light continued to glare red, and the warning horn continued to blow after the gear handle was placed in the down position. Recycling the gear handle had no effect. All through another circuit of the GCA pattern all efforts to get the gear down with sharp jerks on the control stick together with rocking the wings and yawing the aircraft were ineffective.

Now, time was becoming a problem. My fuel state wasn't critical yet. However, other aircraft in the stack were beginning to express needs to get on the ground. Finally, I suggested to Approach Control that they send me back to the top of the stack, and work me back down to about 3000 ft. Then, clear out everyone below me and let me dive the T-6 to about 1500 ft for a 4+ g pullout with the gear handle in the gear-down position. After a small amount of discussion, Approach Control agreed with the plan. The maneuver worked. At the bottom of the pullout there was a loud cracking sound, much like a tree limb breaking, followed by an indication that the gear was locked in the down position.

The post-landing examination showed that when I left Chicago water splashed up into the gear well then froze after the gear had been retracted. The ice was strong enough to overcome the hydraulic system, but not the combination of the hydraulic system and the "g" loads.

ICY C-45

I don't recall why we changed from T-6s to C-45s, but about halfway through our University of Michigan assignment, we did. The C-45 is still around in various modified forms. In the "50s," it was a twin-engine tail dragger that carried about four passengers and a two-pilot crew. Two pilots were not required. Regulations allowed flying with only one pilot.

C-45

Three flights made flying the C-45 at Selfridge memorable. The first flight involved ice. For some reason we had flown to Oscoda Airfield. Oscoda is located about one hundred miles north of Detroit. When the time came to go back to Selfridge, freezing drizzle was coating the C-45 with ice. As usual, we needed to get back to the university for classes. In addition, spending the night at Oscoda was not something we were keen on. There was no hanger space for housing the aircraft while we worked on removing the ice. Ice had already covered the topside of the fuselage, wings, and empennage. There was a wheeled heater available. The heater worked O.K. However, when we moved from one side of the airplane to the other, ice reformed on the ice-free side while we were clearing the ice from the second side. Finally, I decided that we would clear ice in a nose to tail direction as rapidly as we could. When it appeared that we were beginning to ice up at about the same rate we were clearing the ice, we would quickly board the aircraft, start the engines, taxi at maximum safe speed to the runway and take off. After getting airborne, we would climb quickly to above the freezing level (wherever that turned out to be) and proceed to Selfridge.

Everything went reasonably well until we attempted the post-takeoff climb to above the freezing level. With the engines giving all they had to give, and the propellers slinging ice into the fuselage, we were doing well to maintain the altitude we had. Fresh out of clever ideas, I'm confident that I silently mentioned that I sure could use some Divine assistance. The good news was that the engine power was sufficient to hold altitude until we reached the southern edge of the freezing drizzle zone. Except for a harder than usual touchdown at Selfridge because I failed to increase touchdown speed enough for the heavier landing weight, the rest of the flight was uneventful.

CAT NAP

The second non-routine flight in the C-45 was the final leg back to Selfridge at the end of a weekend cross-country. A fellow student and I were taking turns sleeping while the "awake" pilot monitored the performance of the autopilot. As usual, the time was close to midnight; and we were both very sleepy. Some time during my watch, I went to sleep. When I finally awoke, I checked the ground for signs of lights. There weren't any. The stars were still there, but there was not a ground light visible in any direction. After positioning the fuel selectors for maximum access to fuel, I woke up the other half of the crew and suggested that we team up to figure out where we were. It turned out that we were about the middle of Lake Michigan, headed into Canada. A really long snooze on my part could have been big trouble. It sure is great to have uninterrupted watch-care, twenty-four hours each day.

TOO MUCH HELP

The third unforgettable flight in the C-45 at Selfridge was a flight that I had planned to make by myself. While I was making out the clearance, a maintenance sergeant asked if he could come along and log his flight time for pay. Of course, I was happy to have him go along. Having flown only a very few two-pilot aircraft for very long during my flying career, I have no liking for two individuals messing with aircraft controls at the same time. Either I am flying the airplane, or the other pilot is flying the airplane. If the other pilot has the airplane, I will do nothing unless he tells me to. I want the same from him when I have the airplane. So, I made clear to the sergeant that he was not to touch any of the controls. He stated that he understood.

During the takeoff roll, as I looked down to check the airspeed before applying back pressure to lift off, the airplane started sinking towards the runway. I immediately pulled back hard on the control column, but not soon enough to avoid beating the runway with both propellers for a distance of about one hundred feet. Fortunately, the airplane didn't stall and slam into the ground. It did, however, shake so hard all of the way around the pattern that I thought we were going to pop rivets. As soon as I got the wheels back on the runway, I shut down both engines. All propeller blades were bent backward about six inches. The sergeant could give no explanation why he pulled up the gear handle. He did remember me telling him not to touch any of the controls.

The next opportunity to fly the C-45 would come in 1953 at Holloman AFB, New Mexico.

10

Korea

6555TH GUIDED MISSILE SQUADRON DUTY

June of 1952 brought graduation from the University of Michigan's Rackham Graduate School. Fitzpatrick and I again made an all out-assault on the USAF Personnel Officer Assignment Office, seeking duty in Korea. Fitz was fortunate. An order had just been circulated allowing officers who had no previous combat experience to volunteer for a Korean tour. He was assigned to an F–84 outfit at Tague, Korea. I wrote letters and made telephone calls trying to convince the authorities that ancient WWII combat duty should not be considered as "combat experience". Their response was to assign me to Headquarters, Air Research and Development Command, Baltimore, Maryland. My plea that headquarters was no place for a first lieutenant must have found a more receptive ear. Our furniture went to Baltimore. Katie, Terry, Steve, and I drove to Patrick AFB, Florida, for duty with the 6556th Guided Missile Squadron.

Three things made the 6 months stay at Patrick endurable:

1. We found a house on the beach to rent; and we enjoyed our very own strip of sand on the water.

2. I was fortunate to have Major Bruce Arnold, one of General Hap Arnold's sons, for a Commanding Officer. He is a real gentleman and a fine officer. After I had been with the squadron for about three months, he made me his executive officer.

3. I was assigned to the 3597th Flying Training Squadron for flying. The

3597th had a T-33, which was used to simulate a Matador missile flying from Cape Canaveral to the Bahama Islands. I simulated a Matador every time there was an opportunity.

Lockheed T-33A "Shooting Star"
USAF Museum

PENTAGON VISIT

Early in December 1952, I decided to risk a long shot and personally take my case for an assignment to Korea to the Pentagon. So, with no orders, I hitched a ride to Andrews and made my way from there to Officer Personnel in the Pentagon. With the assistance of a captain who appeared to have a free moment, I found my way to a major in personnel who was willing to listen to me.

My message was simple. Either I get an assignment to a fighter outfit in Korea; or, I resign my commission. Clearly, the captain and major had never been confronted with this sort of thing before. There were at least half a dozen responses they could have given me, most of them unpleasant. Happily, they chose to be supportive. Nothing was promised. But they said they would see what they could do.

Very shortly after I returned to Patrick, Major Arnold called me into his office for a discussion about the trip to the Pentagon. General Richardson, the Center Commander, had been notified by the Pentagon about my visit. My standing with him was several notches down the pole from the "unfavorable mark". I received a letter of reprimand for my behavior. A supportive letter from Major Arnold to the General saved me from something worse.

FINALLY, KOREA

In January 1953, orders arrived at Patrick assigning me to the 8th Fighter-Bomber Wing at Suwon, Korea, by way of Nellis AFB to check out in F-86s. At Nellis, we flew F-86As and F-86Es. The F-86A had a 37 ft 1 in wingspan. The F-86E wingspan was one inch shorter. The F86A was 37 ft 6 in long. The F-86E length was 37 ft. The F-86A empty weight was 10,093 lbs. The empty F-86E was 462 lbs heavier. The General Electric J-47 engine in the F-86A produced 5,200 lbs of thrust. The J-47 version in the F-86E delivered 5,970 lbs of thrust. These differences were not particularly significant. There were, however, two modifications incorporated into the F-86E that did make a noticeable difference in the aircraft's handling qualities.

THE 80TH SQUADRON F-86 I FLEW IN KOREA

The first modification was the replacement of the conventional F-86A fixed stabilizer and elevator with an all-moving slab tail. Diving the F-86A to sonic speed resulted in loss of pitch control until the aircraft was slowed below Mach 1. The slab tail eliminated this problem.

The second modification was the replacement of the aerodynamically actuated leading edge slats with a fixed leading edge. Before this was done, undesirable yaw could be produced in a turn by differential slat deployment.

The stay at Nellis was short but sweet. The F-86 proved to be an outstanding machine, a jet P-51. It did everything well. It had no bad qualities. To become combat qualified in the F-86, we did all of the fun things: air-to-air gunnery, ground gunnery, air-to-ground rockets, and angle dive bombing. Ivan Kinchloe was my checkout instructor. Ivan was one of the Korean War aces. Later he flew the X-2 and was killed at Edwards in an F-104 accident.

Katie was pregnant with Tim at the time. She endured the challenging road trips from Florida to Nellis and from Nellis to Little Rock like a champ. Some of the roads in Arkansas would have resulted in an early delivery for most women. She set up housekeeping in half of a duplex next door to her father and mother. I caught a plane for Travis AFB to wait for transportation to Korea, finally arriving at K-13 in June 1953. I managed to get in twenty-two combat missions by the end of July. Then, the war was called off.

Except for the fact that there were lots of Koreans and some hills around, Suwon could have passed for Stuttgart, Arkansas. There were lots of rice paddies. The airfield, K-13, had one long asphalt runway running north and south. The 51st Fighter Group occupied the west side of the field. My outfit, the 8th Fighter Bomber Group, parked their F-86s on the east side of the runway. The 51st Group F-86s had checker tails. The vertical tails of the 8th Group's F-86s had sunburst stripes. The 35th Squadron F-86s had blue stripes. The 36th Squadron F-86s had red stripes. The 80th Squadron F-86s had yellow stripes. I was assigned to the 80th Squadron. The 80th squadron's history went back to WWII duty in the South Pacific. In those days, the outfit was equipped with P-38s.

F-86 COMBAT FLYING

Things were probably different when the 27th Fighter Group, the outfit I left to go to the University of Michigan, arrived in Korea back in 1950. By the time I reported in to the 80th Squadron, combat in Korea was very different from WWII combat. The high price of airplanes had modified the rules of engagement for fighter-bombers. (P-51s cost only $50,000 a copy during the '40s.) In Korea, bombs were released at about 3000 ft. Pilots were instructed to strafe convoys of multiple vehicles, but not to risk aircraft damage by going after a single vehicle. Rules were not always remembered.

During my one month of combat, the 8th Group lost no aircraft to enemy action. In a August of 1953, I became the squadron operations officer. Over the next ten months of post-war duty, three 80th Squadron F-86s were destroyed in accidents. One pilot was killed. The other two were not seriously injured. We did have a fourth pilot killed by a landing F-86 with one gear locked in the retracted position. Three of us were standing near the runway to watch the happening. When the F-86 pilot could no longer hold the wing with the retracted gear up, the wing tip hit the ground. The F-86 veered off the runway towards us. Two of us ran laterally. The pilot that was killed ran in the direction the F-86 was traveling and was run over.

DROPPING THE "RAG" MISHAP

No aircraft accident is funny, particularly to the pilot involved. But sometimes there can be a little humor associated with a mishap. I watched an accident happen off the end of the K-13 runway that brought a chuckle, *after the determination that the pilot had come through unscathed.*

Air-to-air gunnery competition was keen within each 8th Fighter Bomber Squadron and between Squadrons. Bullets were tipped with different color paint to identify the shooter. The goal was to put large numbers of holes less than two inches long in the banner being towed by a long cable. The cable was held in place on the tow F-86 by closed speed brakes. Holes longer than two inches in the banner made the shooter very unpopular with the tow Pilot. It is possible to hit the tow aircraft if the shooter's trigger is held down too long while in the pursuit curve.

Tow speed was important to the shooter. If the banner was towed too fast, the trailing edge whipping action could shorten the banner by a couple of feet. Most shooters were sure that they had lost credit for many hits when the banner was returned shorter than it left. Therefore, conscientious tow pilots tried to keep their tow speeds as low as practical, particularly when they were making the low pass down the runway to drop the rag in full view of the shooters, who were now waiting on the ground.

This particular day, I just happened to be standing near the runway when another squadron's tow plane made the rag dropping pass at about two hundred feet. It was very apparent that this pilot was really holding the speed down. As the tow plane reached the drop point, the dive brakes came out. The cable released. The F-86 stalled. And the tow-plane hit the ground in the scrub trees off the end of the runway. Not funny at all at the moment. But after the pilot climbed out of the cockpit unharmed, it did bring a bit of a smile to one's face.

SHUTTING DOWN THE KOREAN WAR

From August 1952 to April 1953, the three 8th Fighter Bomber Wing squadron's sat a lot of alert, shot a lot of gunnery, and did a lot of dog fighting. When the agreement to call off the war was finally reached at Panmunjom, we received orders to get as many F-86s out of Korea as we could before the agreement went into effect. I think there was to be something like a percentage decrease of the F-86s in Korea at the time the agreement went into effect. We could bring any F-86s that were presently in Tusiki, Japan for maintenance without penalty. Whatever the deal was, the authorities wanted us in Japan that day. The weather was down at Tusiki, as I led the 80th Squadron in a four-flight formation to Japan.

On arrival at Tusiki, we made instrument approaches in flights of four F-86s each. The flights were spaced about five minutes apart. The 80th squadron was the only one that made it. The other two squadrons spent the night at K-8 in Korea.

VISIT TO VIETNAM

In April 1953, I received orders to report to Holloman AFB, New Mexico. While waiting for travel orders, I was offered the opportunity to fly co-pilot on a Group C-47 that was being given to the French in Vietnam. This sounded like an interesting way to wait for a ride to the states. So, I assembled a week's supply of clean clothes and toilet articles; and we were off for Clark Field in the Philippines. The stop at Clark Field was necessary for replacing the C-47s American markings with French markings. This took a couple of days.

We took off from Clark Field with orders to take the C-47 to Hanoi. About halfway between Clark Field and Hanoi, we received a radio call advising us to divert to Tourane (Danang). The French were no longer in charge in Hanoi. I was happy that radio transmissions were coming through well that day. With French markings on the C-47, we would have been hard pressed to convince the Vietnamese that we were Americans. Since Harry Truman was President then, the U.S. Government would probably have admitted that we had been sent. Any time after Harry left office, we would most likely have been on our own.

Tourane was an interesting airfield. The French had quite a collection of WWII aircraft. There were several Douglas B-26s. They even had some Navy Hellcats. What they didn't have handy was a drink of water. A young French pilot and I searched everywhere for potable water. Neither he nor any of his comrades understood why water was so important. There was plenty of wine to be had just about anywhere on the airfield.

There was a prisoner of war compound on the airfield. The size of the prisoners was a surprise. The Vietnamese were such tiny people. They were guarded by large Senegalese blacks. Even more surprising was the discovery that the French were, in effect, prisoners on their airfield. They would visit the nearby village in armed groups during the daytime. No one ventured outside the airfield barbed wire fences at night. The young French pilot spoke English fluently. He told me that all of the surrounding territory was filled with Vietnamese guerrillas.

By the time we made our way back to Korea, my travel orders were waiting. When I arrived at Adams Field in Little Rock, Katie was waiting for me at the airport with my first opportunity to get acquainted with Tim. He was one handsome, healthy, boy. It sure was a joy to be back with Katie and the boys. We weren't sure what to expect at Holloman Air Force Base. We had seen some of the desert at Las Vegas. It wasn't very impressive. As things turned out, Holloman was probably the assignment we enjoyed the most of all the ones we had.

11

Holloman AFB

HOLLOMAN AFB DUTY...1953 – 1956

Holloman Air Force Base is located about five miles west of Alamogordo, New Mexico. In the 1950s, the base occupied approximately ten square miles of space on the eastern edge of the White Sands Proving Grounds, midway between the south and north Proving Grounds boundaries. The mission of the Air Force personnel located at Holloman was to develop missiles, both air launched and ground launched. Actually, Holloman Air Force Base was one element of the Air Force Missile Development Center, commanded by General Leighton I. Davis. From a range operations standpoint, the Air Force was a user of the Army controlled, White Sands Proving Ground.

From a flying standpoint, Holloman was just about as good as it gets for peacetime flying. At Holloman, I was current in L-20s, C-45s, U-3as, B-26s, T-33s, P-80s, F-94Bs, F-94Cs, F-86s, F-100s, and F-104s, all at the same time. I also flew a P-80 with movable vertical fins on the wings, which was a cruise missile flying simulator. Another P-80 that I flew was modified with a huge gyro. The outer ring was about 2 ft in diameter. It was mounted just behind the cockpit. This P-80 also simulated a cruise missile.

To make things even better, our Center Commander, General Leighton I. Davis, was a P-26 fighter pilot in his years as a second lieutenant. He understood how to maintain professionalism in a flying organization without resorting to a maze of petty regulations in a fruitless attempt to do so. He would not tolerate dumb or foolish flying. But, he had no problem with letting Joe Kittinger and I fly over the parade ground wingtip-to-wingtip at about 100 ft and do Immelmans off the deck as the troops passed in review.

Kittinger first became famous by parachuting from a balloon launched near Holloman. The bailout altitude was above 80,000 ft. I believe this record still stands. Joe was also a pioneer in zero "g' aircraft research. He and a medic did considerable zero "g" flying at Holloman in a modified F-94C . Later, Joe was shot down in Vietnam, and spent a short tour in the Hanoi Hilton. After retiring from the Air Force, he sat a distance record with a solo balloon flight from the United States to Italy.

MISSION CONTROL

Another Command Policy that made Holloman such an outstanding place to be assigned was the practice of using flying officers in dual roles. I had the good fortune to be Chief of Mission Control, the operations center from which Air Force test missions were conducted. At the same time, I flew as a test operations support pilot and as a test project pilot on several test and development programs. Quite a few rated officers at Holloman had such dual assignments.

Our missile development tests were scheduled weekly at a joint meeting of the Army, Air Force, and Navy users. Day to day support coordination and changes to the Air Force missions were done from Air Force Mission Control. Mission Control was located in King One, the radar station at Holloman. A hard-working, capable, charger named Dick Shoulders managed King One. An equally capable gentleman named Reggie O'Neil handled the optical and telemetry data gathering and processing net.

The radar plotting boards and ground to air communications links occupied the center part of the building. The optics and telemetry control were done from a large room on the east side of King One. The Mission Control Room was located on the west side of King One. Glass walls partitioned off the three rooms.

Most projects provided their own aircraft vectoring and test control team. More often than not, several tests would be operating on the range at the same time. The direction required to ensure that this was done effectively and safely was exercised by Mission Control, coordinating with the Army White Sands Proving Ground Control Center at Las Cruces. Knowing the positions of all aircraft, and other vehicles (missiles, projectiles, etc.) in the White Sands airspace at any instant of time was a must. By voice-radio, aircraft checked in with Mission Control coming into the range airspace and going out of the range airspace.

Occasionally, high-spirited pilots would buzz King One departing the range to verify that they had indeed departed the range. Wayne Kunkel, a very capable Holloman support pilot and good friend, could definitely be described as high-spirited. I recall that on one day he knocked off the tip of the horizontal tail of a T-33 I was flying with his wing tip while we were practicing formation. Neither of us knew it happened until we parked our aircraft on the ramp by the main maintenance hanger. Always a quick thinker, Wayne checked the aircraft in the maintenance hanger...found a T-33 that was being worked on...and had one of the maintenance troops transfer the horizontal tail tip from that aircraft to the T-33 on the ramp. Beats filing an incident report any day.

On another day, Wayne was again flying a T-33. This time he was on a photography mission and was carrying a photographer in the back seat to photograph whatever test they were covering. I forget whether Wayne decided to buzz King One as he cleared the range or whether a suggestion was made over the radio voice-link that he do so. In any event, Wayne rattled the windows as he came by quite low.

Lest he get an over enlarged ego, the standard transmission for such a performance went out from King One, "When are you going to get off of oxygen?" To which, Wayne replied, "Watch this." Within a matter of thirty seconds, there was a sound from the King One roof like two automobiles having a fender bender at about 80 miles per hour. This was followed immediately with a call from Wayne that he had an emergency situation to deal with. One of his wings struck an antenna on the King One roof, creating a huge gash in the wing. He, and the photographer with him were very fortunate that the aircraft didn't go in.

Needless to say, the happening captured the top priority attention of the Aircraft Operations Chiefs, the Flight Safety troops, and others in positions of responsibility. My report stated that I had instituted the practice of having aircraft buzz King One to confirm that an aircraft was clearing the range. Not a completely factual statement, but very important to the outcome of Wayne's case. I received a letter of reprimand from General Davis, as did Wayne. The letters of reprimand were actually a great favor from the General. I am sure he knew exactly how everything happened. The letters of reprimand were the kinds which are removed from an officer's file when he transfers to another outfit.

About two weeks later, General Davis awarded me an Air Force Commendation Medal for doing something. He smiled when I asked him if I was even with the world again.

From my standpoint, probably the most important piece of data relating to King One was geographical. The radar station was approximately five minutes from the aircraft flight line by motor scooter. I could respond to a requirement for a support pilot in about 5 1/2 minutes. The thirty seconds was the time required to exit King One and get my scooter rolling.

One of the major uses of the L-20s at Holloman was to track balloons. The balloon outfit based at Holloman launched balloons for several purposes. One of those involved carrying payloads, some weighing several hundred pounds, across the United States wherever the wind carried them. The payloads were ultimately parachuted to the ground.

An L-20 was sent along to monitor the happenings to the balloon and help locate the balloon and payload when they hit the ground, intentionally or unintentionally. Often, the balloons drifted over territory where there were no airfields. So, the L-20 pilot was authorized to land in a field, on a road, or wherever he could. One such road landing by a Holloman pilot attracted several curious motorists who stuck around to watch the takeoff. For whatever reason, the pilot chose to take off in the direction of a power line which crossed the takeoff highway, rather than take off in the opposite direction. The crowd was treated to front row seats for the "Crash of the Day" when he hit the power line. Fortunately, he was not seriously hurt. But the airplane was a Class-26 (not repairable) case. Another Holloman support pilot flew an L-20 up a box canyon and didn't survive.

I checked out in the L-20 to be qualified for balloon chase, but never flew a balloon chase mission. The only support mission that I flew in an L-20 was to photograph an F-102 as the aircraft launched a missile. There were no chase aircraft available that day that could fly formation with the F-102. So, with a photographer on board, I flew a tight circle at the point over the range that the launch was to be made at an altitude of 10,000 ft. Listening to the launch countdown from King One, and watching the direction from which the F-102 was coming, I timed my circle to have the photographer in position for a shot out the right window of the L-20 as the missile left the rail.

DeHavilland U-6A "Beaver"   USAF Museum

L-20

The L-20 was also used to transport individuals to various dirt airstrips on the range. This was always a fun thing to do.

FLYING THE C-45 AT HOLLOMAN

At Holloman, the C-45 was a pilot proficiency and personnel transport aircraft. The rocky airstrip at the north of the range was narrow, rough, and short.

But the C-45 could make it in and out. The Range Safety Officers and other support personnel made the round trip many times. Two C-45 flights at Holloman are etched in my fading memory.

C-45

WHERE TO "GO" AND HOW TO "DO IT" QUESTION

The first happening in the C-45 was a local night flight that I made to log time. Very rarely did I need to make a flight to log four hours for flight pay. But this was one of those times. No one else was interested in getting airborne that night. So, I was flying alone. I shot landings for a while. Then, I visited the east, west, north, and south extremities of the range a time or two to lessen the boredom. Suddenly boredom was replaced by intensifying stress as the need to urinate became more and more critical.

In the C-45 there is no relief tube in the cockpit for the pilot. The "John" is in the tail end of the airplane. Letting the autopilot fly the airplane for a couple of minutes crossed my mind. However, I remembered times when the autopilot suddenly made the C-45, roll, pitch, dive, or do a combination of the three. The autopilot could not be trusted.

I really didn't want to land the airplane, get out and relieve myself, and get airborne again. But I was just about to do it, when I noticed some Sectional Charts in the co-pilot's seat. A perfect solution! I made a cup from one of the Sectional Charts and filled it to the brim. The C-45 had a bad habit of rolling left or right when the control wheel was left unattended. So I kept the wings level by yawing the airplane with the rudders. Holding the cup with a steady right hand I opened the window on the left side of the cockpit with my left hand. Very slowly and carefully, I brought the cup to a point just forward of the front edge of the window opening. The final step was to be a quick toss of the cup out the window. I managed to hit the hole squarely with the cup. Had the air flow by and through the window been more favorable, I would have been a happy troop. As it turned out, the air flow was most unfavorable. I was one wet, smelly, aviator. Things would have been much better if I had simply urinated in my flying suit. Some lessons are costly!

BILLBOARD READING AT NIGHT IN A SNOWSTORM

The second memorable C-45 flight was a weekend cross-country. Several airmen from Holloman were making the round robin with me. The airman riding in the co-pilot's seat was one of two young men assigned to Holloman as corporals. In college, they had completed the ROTC requirements for a second lieutenant's commission. However, they had chosen not to commit to two years of active duty as officers. The penalty for not doing so was to be placed on active duty as corporals. I never quite understood how all of this came about. But, it did. They were two very capable young men. A few months later, they were commissioned anyway.

I made one of my typical round robins...fly the entire time, stopping only to eat, go to the "John", and catch an occasionally nap. It was wintertime everywhere. There was a fair amount of winter weather around when we headed back to

Holloman about midnight on Sunday. As we passed Oklahoma City, the weather prophet advised that ceilings in the El Paso/Holloman area were very low, and it was snowing. El Paso Municipal was unquestionably the best place to go. We could nap somewhere in the El Paso terminal until daylight, and fly to Holloman when conditions were more favorable the next day. The temperature was something well below freezing.

My available navigation radio equipment dictated a VOR approach, which was O.K. The ceiling was reported to be around 700 ft. Had it not been snowing so hard, it would have been a piece of cake. Even so, I didn't anticipate any great difficulty. Normally there would have been none.

Generally, a person can handle one difficult problem. Things get pressing when one difficult situation couples with one or more unexpected ones. The unexpected problem was something that I had never before experienced. Because of the low temperature, the oil in the right engine propeller control cylinder was intermittently congealing. As a result, the propeller began to cycle rapidly back and forth from low to high pitch. This caused the aircraft to rapidly yaw hard right then hard left. The result was that I focused all of my attention on trying to control the changing yaw and to get the propeller pitch stabilized. All of this happened in a minute or less. As I mentioned earlier, the C-45 does not wish to keep its wings level when the pilot leaves it unattended. The aircraft behaves like it is balanced on a knife-edge. You soon learn that when you are flying instruments in a C-45 you check your attitude indicator very frequently. Even though I had been conditioned to do this, I am certain that Divine providence was again working overtime when a realization flashed through my mind that too much time had passed since I had checked my attitude. Instantly, I checked. The aircraft was in a ninety-degree bank, going down. Remembering that I was somewhere between one foot and 800 ft, I rolled the wings level and pulled back hard on the control column. Through the snow, for just an instant, I saw a highway billboard illuminated by my landing lights as it disappeared a few feet below the nose of the aircraft.

I let the propeller cycle and the aircraft yaw while I climbed back up, found the VOR inbound approach leg again, and bracketed it to a landing. For sure, there was only one set of footprints in the snow at El Paso that night.

FLYING THE U3A (BLUE CANOE)

The U3A (L-27) was a twin-engine Cessna that carried four or five people at a cruise speed of about 180 kts. Because of its paint job, most of the Holloman pilots referred to the U3A as the "Blue Canoe". I logged about 15 first pilot hours in the U3A. All of them were routine cross-countries, proficiency flights, or personnel transportation hops around the White sands Test Range. The aircraft was easy to fly and land. It handled satisfactorily on one engine and had no bad stall characteristics.

Cessna U-3A

I recall riding co-pilot with Dick Corbett, a good friend and fine pilot, on what would have been an uneventful flight except for the demonstration he performed for me. We were preparing to land at Holloman when Dick asked me if I would like to see a touchdown and stop within five hundred feet of the approach end of the runway from a point five hundred feet above the approach end of the runway. Since he, not me, would get credit for any miscue, I stated that I would like to see this done. With gear and land flaps down, Dick slowed the "Blue Canoe" to just a few knots above stall speed and maneuvered to the intended start point. At the instant we arrived over the end of the runway, Dick chopped both throttles, stalled the airplane, and dumped the nose. The stall recovery was the landing flare. There was room to spare within the predicted 500 ft landing roll limit. I smiled to myself as I recalled demonstrating this technique in a Stearman during Primary Flight Training to my instructor, Mr. Grey.

About half of my 15 hours in the U3A was accumulated going to and from Ogden, Utah. Hal Ebersole, a skilled project pilot and good friend, and I were investigating an accident involving a Holloman F-100 crash at Ogden. A Holloman support pilot named Cohn was killed bailing out of an F-100 shortly after takeoff. Cohn experienced an explosion in the bottom, rear part of the fuselage engine compartment during gear retraction. He knew exactly what was happening. These explosions were occurring in F-100s throughout the Air Force. Fuel to the afterburner fuel spray-bar in the tailpipe of the engine was delivered via short, curled tubes (pigtails) encircling the external surface of the tailpipe. Engine vibration was producing fatigue cracks in these pigtails, spraying raw fuel between the fuselage skin and the tailpipe. This fuel would puddle underneath the tailpipe. Heat would vaporize the fuel, creating a bomb in the aircraft's tail-end. A spark did the rest. Unfortunately, the slab horizontal tail positioning actuator was co-located with the bomb. About 30 seconds after the explosion, the slab tail would lock up. Knowing that this sequence was in progress, Cohn gained as much altitude as he could before lockup (about one thousand feet) and pulled the ejection handles.

Cohn did everything exactly as he should have, according to the book. As it turned out, a piece of survival equipment that was meant to save the pilot killed him. In the F-100, the pilot sat on a survival kit containing "things" to help him survive on the ground or in the water. According to the book, the survival kit was to be attached to the parachute harness as part of the pre-engine start cockpit check. Unless I was planning to fly over large bodies of water, I always chose to ignore that part of the book and leave the survival kit unattached.

Cohn's ejection seat and the automatic seat release worked exactly as advertised. He had a clean separation from the seat. The pilot chute came out as it was designed to do. Unfortunately, the survival kit straps, which attached the kit to the parachute harness, were not cinched up. The pilot chute went between Cohn's rear end and the survival kit. When he straightened up, the pilot chute was caught, and the main chute canopy made a "horseshoe." As a result, he was killed when he landed in a freshly plowed beet field near the airport.

REPEAT F-100 ACCIDENT AT HOLLOMAN

Not too long after Cohn's bailout, Capt. Buck Buchanan, a long-time friend and fellow traveler, had a repeat of Cohn's accident immediately after an F-100 take off at Holloman Air Force Base. Buck did all of the right things. His parachute brought him down safely.

FLYING THE DOUGLAS B-26

Another Douglas winner!!!!! Truly, the B-26 was another choice chip off of the block that the A-20 came from. It did everything well. Somehow, it never occurred to me to see if the B-26 would roll and loop like the A-20. I'm confident that it would have. It had a high cruise speed. Handling qualities were excellent. Landing was a piece of cake. It was solid as a rock on instruments. Like the A-20, the B-26 was a terrific low-level machine.

Douglas A-26A "Counter-Invader"   USAF Museum

At Holloman, the B-26s had several uses. They were used for test parachute drops, Ryan Q-2 jet drone drops, and other tests. It also served as a rapid response cargo and personnel carrier. Most of my B-26 flying was in the cargo and personnel carrying use category. I did chase and serve as safety shoot-down aircraft for Q-2s that were launched from B-26s.

Three flights come to mind when I check the memory bank for B-26 recollections. The first was a midnight B-26 flight with Major John Pitts returning from Andrews AFB, near Washington D.C., to Holloman AFB.

WHOSE TURN IS IT TO FLY?

In addition to being an excellent pilot, John was a fine officer and a gentleman in every sense. He enjoyed life and made life enjoyable for those around him. He was eventually promoted to brigadier general.

By the time we began our last leg into Holloman we were both very tired and sleepy. So, we did the right thing by agreeing to take turns flying. Practical jokes were not unheard of in our relationship. It occurred to me that this would be a good opportunity to square a score or two. I took the first piloting turn. In a very few minutes, John was sleeping soundly. After about fifteen minutes, I set the aircraft clock forward by forty-five minutes and awakened John for his one-hour turn. An hour later, John shook me awake. This time I let him sleep for about twenty-five minutes before setting the clock forward and welcoming him back to the world of the wide awake. He complained that he felt as if he had hardly slept. Without responding, I promptly began my-one hour snooze. When my sleep hour was up, we were only about thirty minutes from Holloman. John dozed a little before we landed.

I said nothing to John or anyone else about our flight until we had a Dining In at the Officer's Club a few weeks later. I asked for permission to speak after the smoking lamp was lit. John's reaction to my account of the last leg home in the B-26 was even more enjoyable than the extra sleep.

"IF ANYONE IS LISTENING, I NEED AN IFR CLEARANCE!!"

The second memorable B-26 flight was a night contest with the weather. Cohn, the officer who was later killed in a F-100 accident, was coming back with me from some place in the southeast about ten o'clock at night in a B-26. I had filed VFR, even though stormy weather was forecast across central Texas. As we left Louisiana and entered Texas, we could see continuous lightning from far south to far north ninety degrees to our flight path.

Clearly, as a minimum, I should change to an instrument flight plan. All of our efforts to contact someone who could authorize a flight plan change did nothing for us. We were still trying as we entered the first thunderstorm. Instantly, survival replaced a flight plan change as the primary consideration.

Douglas built their flying machines to be sturdy. This poor airplane shuddered and bucked constantly as the view out the forward windscreen went from pitch black to blinding light. My attention was totally locked in on keeping the airplane upright and headed in a westerly direction. Occasionally I checked the altitude. The altitude readings varied considerably. We went through some heavy rain. However, it wasn't the "flying through Niagara Falls" downpour that I had experienced in other thunderstorms.

Mercifully, we made it through the squall line in about twenty minutes. The rest of the flight to Holloman was smooth and uneventful in VFR weather conditions. It turned out that a flight plan change to IFR didn't really matter all that much, anyway.

SEARCHING FOR THE RUNWAY IN A B-26

Night weather was the attention getter in the third special B-26 flight that I recall. Some time around midnight I was returning to Holloman alone from up east somewhere. I was at the end of a long leg, and quite low on fuel. The plan was to land at St. Louis Municipal for refueling. When I arrived at St. Louis, that part of the country was experiencing a very heavy downpour. From whatever the approach options were, I had chosen to make a GCA.

The GCA controller was superb, a real professional. Each time he would bring me around the pattern quickly and advise that I was over the threshold and should take over for a visual landing. Three times he did this. Three times all I could see through the blinding rain was a fuzzy glare from the approach lights. Checking the fuel remaining, I estimated that I could fly the few miles to Scott AFB for one approach to a landing there or stay at St. Louis Municipal for two more approaches. Since there was no reason to believe that the weather would be any better at Scott, I decided to stay where I was. The fourth try at St. Louis was a repeat of the other three.

Sheets of rain were still coming down, as I came around the pattern for the fifth time. Checking my fuel state, I decided that if the controller puts me over the threshold this time, there will be some sort of landing. When the controller said, " You're over the threshold," I chopped both throttles and applied back pressure to slow my rate of descent. As the main wheels touched, I had my first opportunity to see a runway light, as it passed directly under the fuselage. The aircraft was straddling the left row of lights. As the left main gear started dragging the fuselage nose to the left, the left wheel sank into the mud. Reaction rather than logic saved the day. Without really thinking, I left the right throttle in idle, and went full forward with the left throttle. Hard right rudder and full power on the left engine were enough to get the airplane back on the runway.

Miraculously, the left propeller did not hit the ground or a runway light. Only one set of footprints in the mud that night.

FLYING THE T-33

The "stretched" P-80 that the Air Force christened the T-33 was the Air Force jet training workhorse for a long, long time. It had its faults; but, overall, it was a great trainer. In my opinion, the only two undesirable features worth mentioning were the cramped, poorly laid out cockpits and the muscle building, lateral control stick forces without aileron boost. Pilots almost never lost aileron boost. So, really, the only annoying pilot task was getting strapped in the cockpit on a hot summer day when the humidity was high. Fortunately, the humidity at Holloman was usually on the dry side.

The T-33 handled very well. Except for an interesting departure which could make the aircraft tumble, cartwheel, or go into some other unconventional maneuver, the spin recovery was normal and very positive. To get into the unconventional mode, the pilot put the aircraft into a conventional spin. Then, while holding the rudder in the direction of the spin and the stick all of the way back, move the stick laterally to the other side of the cockpit away from the spin direction. To recover, neutralize the controls and wait. Recovery could require several thousand feet.

T-33s were the primary training and pilot proficiency jet aircraft at Holloman, as they were at most Air Force bases. They were extensively used for combination flight proficiency and administrative travel flights. They were also used as chase and photography aircraft.

I may have had other out of the ordinary T-33 flights at Holloman, but three come to mind now. Two of the flights were made at night with "Buck" Buchanan, the friend who bailed out of the F-100.

Buck was flying in the front seat on the first of the two flights. As the story so often begins, "It was a dark and stormy night." The time was about midnight. We were returning to Holloman from some place on the west coast. A line of very tall thunderstorms from the north end to the south end of the San Andrea Mountains blocked our way for a VFR approach. We were too low on fuel to chance an instrument letdown and approach to Holloman. However, El Paso was within easy reach and the weather there was reported to be VFR. Just as we were about to turn for El Paso, we noticed what appeared to be possible VFR paths between the thunderstorms which became visible when the lightning flashed. So, we agreed that I would be the "path spotter" and call heading changes when a lightning flash occurred. Buck would fly the aircraft. We were doing reasonably well until the time between lightning flashes became longer. Suddenly, we found ourselves IFR in a thunderstorm. Reversing course, we flew west until we could see the stars again and proceeded VFR to El Paso Municipal.

T-33 MIDNIGHT LANDING ON A DIRT FIELD

Aviation history records many undesirable endings to flights which had their genesis in a pilot's cavalier attitude about planning for the unexpected. Buck and I barely escaped paying the piper for my inattention to possibilities on another dark, but not stormy, night. We were again returning from the west coast about midnight in a T-33. This time, I was in the front cockpit. I had flown this route so many times that I needed no map or radio aid to navigate to Holloman. Furthermore, I knew to within a few hundred pounds how much fuel was required to make the trip at 35,000 ft or on the deck. At high altitude, the trip could be made without using fuel from the wing leading edge tanks. Following the terrain contour emptied the tip tanks, main wing tanks, and most of the wing leading edge tanks. After takeoff, for no good reason, I suggested that we make the trip home at low altitude rather than do "zero challenge" navigation at 35,000 ft. Buck expressed no preference. So, we stayed low. When the sky is clear and the visibility is unlimited, night navigation over the southwest desert is a piece of cake. There are so few lighted towns and cities that every one you see is easily identified. By the time we reached a point from which the town of Truth or

Consequences could be seen, the tip tanks were empty, and the main tanks were close to dry. Fuel consumption had been exactly as expected. Within less than an hour, we should be in the sack catching up on sleep.

When I placed the wing leading edge fuel tank switches in the "ON" position, cockpit lighting took on a distinctive red hue, signaling that some abnormal condition needed immediate attention. Logically, the first thing to check is the system affected by the control input you had just made. A check of the fuel control panel revealed that fuel pumps for the wing leading edge tanks were not operating. Circuit breaker checks and switch recycling had no effect. The only accessible fuel remaining was that in the fuselage tank. This was not enough to get us to Holloman or El Paso.

I had never looked at the dirt field at Truth or Consequences. Furthermore, I had no idea how long the field was. However, when you are down to two choices, land on what is available or bail out, I prefer landing. So, after one low pass with the landing light illuminating the terrain, I made a rectangular pattern and used power to make a low speed, flat approach that just cleared the wire fence at the south end of the airfield. As the fence passed under the nose, I chopped the throttle, touched down, and rode the brakes to a stop. I had the distinction of making the only jet landing ever made at Truth or Consequences. This noteworthy accomplishment in no way diminished the ill will and critical remarks from the higher echelons at Holloman regarding my planning and execution of the night flight from the West Coast.

WHO'S GOT IT!!!!

My third non-routine T-33 flight at Holloman was made "out of the ordinary" by failure to communicate. Probably fifty percent or more of the pilots who have shared the piloting task with another throttle jockey have experienced an incident similar to the one I had.

I'm not completely positive. But I believe I was in the rear cockpit. I also don't recall the reason that we were airborne. It was probably an instrument-training ride. I do remember that when we finished our primary mission, the gent in the other cockpit suggested that we spend some time flying for fun. That sounded like a great idea, and I told him so. I expected a couple of rolls, an Immelman or something of the sort. All that happened was the airplane slowly spiraled to the left into an almost vertical dive.

After diving for several thousand feet, the T-33 reached an impressive airspeed and smoothly pulled up into a vertical climb. This time I thought we will see a vertical roll or two, an Immelman, or perhaps a Cuban eight. Not so. The aircraft continued straight up until the vertical speed was zero, did a tail slide, and then flopped over into a nose down attitude. After picking up flight speed, the aircraft began a shallow spiral towards the desert floor. Finally, the T-33 settled down into a straight course descent at about a thousand feet per minute. As we passed through 1000 feet above the ground, I began to wonder just how low does this guy intend to go. At about 100 feet above the ground, I was two seconds away from taking control and stopping the descent when he pulled the nose up. He immediately voiced an apology over the intercom for taking control from me. I'm confident that thousands of pilots can tell a true story very much like this one. There is no acceptable substitute for clear, complete, verified communications in the flying game!

FLYING THE P-80 ....CRUISE MISSILE SIMULATORS

Most of the P-80s at Holloman were drones. These were not flown by the Holloman support-pilots or pilots like myself. We flew support-airplanes and project-airplanes. A drone squadron stationed at Holloman flew both B-17 drones and P-80 drones.

Lockheed F-80 C "Shooting Star"
USAF Museum

LOCKHEED P-80

I did have the good fortune to fly two modified P-80s. One was used as a cruise missile simulator for the F-80/B-50 project. This aircraft had vertical fins on the wings. The fins were controllable to produce lateral aircraft motion without banking. When on autopilot, a ground control team could fly the aircraft as they intended to fly the cruise missile launched from a B-50. Actually, I only recall flying this airplane once or twice. It was an interesting ride with no surprises.

CRUISE MISSILE TERMINAL GUIDANCE SYSTEM TESTS

The other project P-80 was assigned to the White Rogers Engineering Co. They were doing research and development on the terminal guidance to target impact for an unmanned cruise missile. The modified P-80 was the flying simulator for this vehicle. The fuselage fuel tank located behind the pilot's seat in a standard P-80 had been replaced with a very large gyroscope. The rotating wheel was approximately twenty inches in diameter. The entire assembly must have weighed a hundred pounds.

With some variation in the autopilot controlled terminal guidance profile, such as entry velocity and altitude, dive angle, and pull-out altitude, the test mission didn't change much. The P-80 pilot, responding to voice commands from King One, flew to a predetermined starting point in the sky. At the instant the aircraft arrived at the starting point, the ground controller said, "ENGAGE". The immediate pilot response was to put the autopilot switch in the "ENGAGE" position and become an observer. One of the things to be observed was a needle that moved across the face of the circular instrument in which it was mounted. The reason for monitoring the needle was to disengage the autopilot just before the autopilot commanded the pitch control to execute a maximum "g" pullout. The pullout needle always had ninety-nine percent of my attention as we approached the "pullout" mark on the dial. One Holloman pilot, a Captain Floyd Kniss, missed the pullout part of the preflight briefing and experienced the autopilot-controlled pullout. The "g" forces tore the wing tip tanks off of the airplane and motivated Floyd to conduct an extensive, unsuccessful search for the party who failed to tell him what to expect at the end of the test maneuver.

Only one of the many test profiles that I flew was unusual. My mishap was the result of a "gain setting" mistake in the guidance system. The flight anomaly that resulted occurred at the very beginning of the terminal guidance profile. The instant I engaged the autopilot in response to the ground command to "ENGAGE", the aircraft did about three snap-rolls before I managed to "DISENGAGE". The post-flight check revealed that one of the gains in the guidance loop was ten times what it should have been.

In retrospect, the most memorable experiences flying the White Rogers P-80 related to getting in the airplane, starting the engine, and operating the radio. There was no ejection seat. In fact, when the pilot showed up for the flight, the seat and the canopy were lying on the ground beside the aircraft. For some reason, it was necessary that the aircraft electrical system have external power hooked up for about an hour before the flight. External power was not obtained from a standard portable power unit. It came via a long, thick cable from somewhere in the hanger. The electrical system could not tolerate an interruption in electrical power. The transfer from the external power cable to internal aircraft power was time limited. When preflight checkout and preparation reached the point for the pilot to man the aircraft, the transition from cable to internal power was made. The seat was installed. The pilot climbed in. The ground crew fastened the canopy to the fuselage. And the pilot started the engine.

Every engine start I made in the White Rogers P-80 was a hot start. Normally, the engine is inspected after a hot start to ensure that no unacceptable damage was done. After a few 1000 deg + starts, we quit commenting about the engine behavior.

The radio was a continuing source of irritation. I recall a couple of flights on which I transmitted by holding a bare wire against a piece of metal in the cockpit and removing it to receive. The White Rogers engineering outfit had to have been on a low, low budget.

FREE 72 HP AIR COOLED ENGINES

Sometimes there were opportunities to fly the P-80 just for fun. Maintenance test hops usually only required thirty or forty minutes. The fuel remaining after this was done could be used for whatever suited you. The Army had assigned an Army sergeant to Mission Control as an Army Liaison Noncom. We worked together very well. We also had a mutual interest in many things. One set of these "things" was 72hp air cooled engines used to power OQ-19 drones which the Army used for Nike Missile tests. We had all sorts of ideas of how these engines could be used.

The Nike testing done at the north end of the White Sands Test Range was near an ancient lava bed. Often the drones would crash or be parachuted into the middle of this lava bed. The lava formation was a maze of crevices, thinly covered bubbles and other hazards. In the summertime the place was infested with rattlesnakes. Travel on foot in the lava bed was very difficult. Travel by vehicle was out of the question.

Having spotted several of the downed OQ-19s, I suggested that we acquire, or make, some pack boards and carry the engines out on our backs. First, we needed to know where to enter the lava beds and how to proceed directly to a downed OQ-19. We finally came up with a solution to the ground navigation problem. The solution required me to fly an aircraft directly from a checkpoint on the west edge of the lava bed to a target OQ-19 and record the compass heading. This would be done for all of the OQ-19s we decided to recover. The best checkpoint turned out to be an abandoned observation tower. The P-80 happened to be the aircraft available on the day that this part of the project was completed.

Later, we drove up to the observation tower in an Army weapons carrier with our tools, pack-boards, and compass heading data. With a strenuous, long day effort, we retrieved five engines. I intended to use my engines for a homebuilt. Eventually, I gave them away rather than move them to my next duty station.

FLYING THE F-94s

There were two F-94 models to fly at Holloman, an F-94B and an F-94C. They were actually two different aircraft designs. The F-94B was a modification of the T-33A, equipped with an afterburning engine. The F-94C was a new interceptor design, powered with a higher thrust engine.

F-94B

The F-94B military thrust at sea level was 4,600 lb. Maximum sea level thrust with afterburner was 6,000 lb. Empty weight was 10,064 lb. Maximum takeoff weight was 16,844 lb. Maximum speed at sea level was 606 mph. The service ceiling was 48,000 ft. The F-94B wing loading at maximum takeoff weight was 72 lb/ft^2.

My guess is that Holloman had the F-94B because no one else wanted it. The aircraft did have an interesting fire control system for delivering unguided rockets in salvo. The feature that made things interesting was how close to the target the fire control system took you. Even the practice passes terrified at least two people, yourself, and the pilot of the target aircraft. Practice was best done in actual weather with zero visibility conditions.

The F-94C was powered with a Pratt & Whitney J48-P-5 engine. Maximum sea level thrust in military power was 6,350 lb. Maximum sea level thrust in afterburner was 8,750 lb. Empty weight was 12,708 lb. Maximum takeoff weight was 24,284 lb. The wing loading at maximum takeoff weight was 72.44 lb/ft^2. The service ceiling was 51,400 ft. The combat ceiling was 49,700 ft.

If the Holloman F-94C had a functioning fire control system, I never used it. Most of my flights in this aircraft were chase missions or maintenance check flights. Several times, I had the opportunity to chase the Lockheed X-7. The X-7 was a ramjet powered, unmanned research vehicle. It looked very much like a "pint sized" F-104.

LOCKHEED F-94C

The X-7 was dropped from a B-50 at 15,000 ft or so. Once the vehicle cleared the launch aircraft, JATO bottles were ignited. These boosted the X-7 to a velocity that would support combustion in the ramjet engine. Since the F-94C was not supersonic in level flight at any altitude, the chase time was very brief. The best we could expect was some photographic footage of the release from the B-50 and part of the JATO bottle boost. The chase technique was to hold a trailing position some distance behind the B-50 and listen to the launch count over the radio. At a point in the count, which had been learned by practice, the throttle was advanced, using afterburner as necessary, to arrive at a point slightly behind the B-50 just as the X-7 was released.

In addition to getting the photographer in the correct position for separation, the F-94C speed and acceleration profile was designed to keep the X-7 in the photographer's view finder as long as possible without having the F-94 pass the X-7. All of this photographing action occurred in a matter of seconds, as the X-7 was boosted to ramjet engine start speeds. Doing it well was always a challenge. The X-7 had an unusual recovery system. Like the F-104, the X-7 had a needle-nose fuselage. Unlike the F-104, the X-7 front end was designed strong enough to withstand a vertical jab in the ground at parachute descent speeds. When the ramjet engine shut down, a parachute deployed from the rear of the vehicle and held it in a nose down vertical attitude. After covering the launch, the F-94 chase would hang around for the X-7 recovery. With assistance from the King One radar tracking crew, the F-94 pilot would locate the landing site and give a verbal description of the condition of the X-7. All of the X-7 landings that I recall were nominal. That is, the "nose spike" would be in the ground. The X-7 would be near vertical. The parachute would be draped over the backend of the vehicle.

F–94C DRAG CHUTE DEPLOYMENT

The F-94C Pilot's Handbook stated that deploying the landing drag chute while 10+ feet above the ground was acceptable. The handbook was correct. The first time I did this, there were a couple of seconds of uncertainty about the outcome. The retarding force was quite noticeable. However, it turned out that the touchdown was smooth and soft.

F–94C DEAD STICK LANDING

Except for aircraft that I flew only a few times, I have practiced dead stick landings in all of the aircraft that I have ever flown. What this does for the pilot is reduce the complete engine failure problem to one of managing the aircraft's total energy at the time of engine failure (potential energy + kinetic energy) to get the aircraft to the landing flare point. Once the pilot has learned the energy cost associated with each maneuver required to get to the flare point, a dead stick landing can become almost routine. Sailplane pilots do dead stick landings all of the time.

I don't recall exactly the altitude at which the fire warning light of an F-94C I was flying near Holloman illuminated. But it was at an altitude that allowed me to shut the engine down without any doubts about having plenty of total energy to make any of the available runways. The fire warning light went out when the engine was shut down. And the landing made into the wind was one of the routine kind. It turned out that the alternator had become hot enough to trigger a heat sensor.

FLYING THE F-86

When I reported into Holloman after the Korean tour, it was a joy to discover that their high-performance, chase-aircraft were F-86As. They were used as shoot-down aircraft for rogue drones. They were also used for chasing Q-2 drones. They were the chase workhorses for just about all tests that didn't require an airborne photographer. Near the end of my first tour at Holloman, an F-86H was added to the stable of support aircraft. The F-86H was a larger and heavier aircraft with a higher thrust engine. For pure flying enjoyment, I preferred the F-86E.

FLYING THE F-86 AT HOLLOMAN

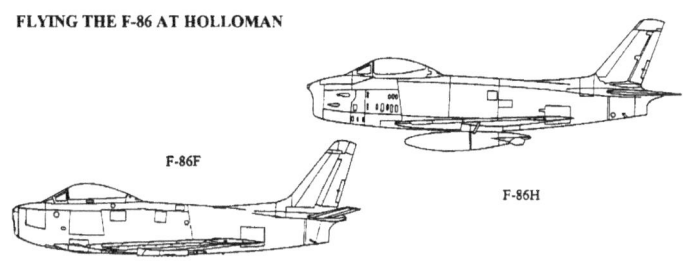

F-86F

F-86H

The F-86H's wingspan measured 39 ft 1 in, two feet longer than the F-86A's wing. Empty weight was 13,836 lb, 3000 lbs heavier than the F-86A. The F-86H was powered by a General Electric J73-GE3. The engine delivered 8,920 lbs of thrust at military power. The F-86H was about 13 mph faster than the F-86A.

I don't recall ever having a serious problem with any of the Holloman F-86s. Our maintenance troops were first class. And the F-86 was a well-built machine. Flight test and flight test support flying were for the most part flown as planned. Pilots had considerable latitude in the kind of flying we did for proficiency. I recall being authorized to put on a single F-86 air-show over the middle of the airfield one afternoon for visiting high school students. No limits were put on what I did or how low the maneuvers were done.

F-86 FORMATION ACROBATICS

The only out of the ordinary happening in an F-86 at Holloman that I recall occurred during a formation proficiency flight with a relatively inexperienced pilot. About thirty minutes into the flight, I suggested that he take the lead and try some formation acrobatic maneuvers. I was impressed by the commendable job he was doing until we started the down side of one of his loops. Suddenly, without warning, he brought his throttle to idle, and deployed his speed brakes. I immediately deployed my speed brakes, retarded the throttle, and looked ahead. It was instantly apparent that survival was the challenge facing both of us. We were on the pull out side of vertical. The only remaining control alternative was to keep the F-86s on the edge of a stall with back pressure, hoping to clear the ground. Fortunately, there was room between us and the desert floor at the bottom of the loop.

HOMETOWN WAKEUP CALL

One very enjoyable F-86 flight was made from St. Louis to Little Rock on a Sunday morning about the time for the locals at Batesville, Arkansas, to leave home for Sunday School. Batesville, the town that I had the good fortune to grow up in, is the best home- town a kid ever had. I was returning to Holloman with some igniters for pyrotechnic devices, which we were using in one or more of the test projects. The F-86 was clean (no drop tanks). So, all acrobatic maneuvers were allowable. A low pass with full throttle at roof top altitude brought most of the folks outside. For the next fifteen minutes, I made my best effort to keep them outside with a low altitude acrobatic demonstration. The performance made the front page of the local newspaper. Credit was given for waking up late sleepers and increasing church attendance.

FLYING THE F-100

I should have kept a diary! I recall going somewhere to pick up an F-100A before I left Holloman to go to Test Pilot School. But I can't recall where. It was probably some place like Tinker Field at Oklahoma City. There was no checkout in the F-100A because the aircraft only had one cockpit. The first flight was the checkout.

Unlike the F-100D, the F-100A and the F-100C had no wing flaps. The "A" model stalled at 159 mph. After George Welch, a company test pilot, was killed doing structural testing of the F-100A, they were grounded until structural modifications were made.

The F-100A had a 38 ft 9 in wingspan. At a maximum takeoff weight of 28,899 lbs. The F-100A wing loading was 75 lb/ft². The F-100A was powered with a Pratt & Whitney J57-P-7/39, two-spool engine. Thrust at military power was 9,700 lbs. Thrust in afterburner was 14,800 lbs.

F-100

The new thing that the F-100A added to my flying experiences was the ability to go supersonic without diving the aircraft. At 35,000 ft, the F-100A could reach Mach 1.4 in level flight. This was something that the early Convair F-102s couldn't do. I recall having to crack the speed brakes to avoid overrunning the early F-102s when flying chase on their tests. Later, the coke bottle fuselage design brought the F-102 into the supersonic class.

12

Test Pilot School

SIX MONTHS AT THE USAF TEST PILOT SHCOOL

G ood news came along in March of 1956 in the form of orders to the USAF Test Pilot School. Katie, Terry, Steve, Tim, and I loaded our belongings and moved to Wherry Housing at Edwards. By this time, we had adapted to the desert. The Edwards climate was about the same as Holloman. We did miss having the New Mexico mountains only a twenty-minute drive away.

Duty at the Test Pilot School was intense. Class 56-C had a total of ten students, eight Air Force pilots and two civilian flight-test engineers. We spent just about every weekday night, Saturday, and Sunday studying, reducing data, and preparing test reports.

The only school aircraft that I had not previously flown were the straight wing F-84, the T-28, and the F-100C. The school also had F-86s, a B-25, and T-33s. These I had flown before. The T-28s, F-86s, F-84, and T-33s were used for performance testing exercises. The B-25 and T-33s were used for stability and control testing exercises. Students only flew the F-100C for two or three qualitative evaluation flights.

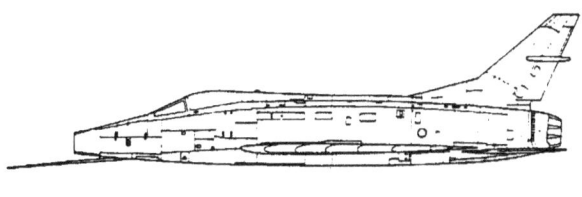

F-100C

The test pilot training was exceptionally well done. We learned a great deal in our six months assignment. I was concerned that they were going to draft me as an instructor. Instructing was not what I was looking for at the moment. Later, I did voluntarily join the Test Pilot School Staff.

TEST PILOT SCHOOL FLIGHT TRAINING

With one exception, all of my flying as a student at the Test Pilot School was according to plan. The one exception was an F-84 performance flight. The School F-84 was equipped with tip tanks. We usually carried fifty gallons in each tank for performance test flights. There was a switch on the left console that, if manipulated in the wrong way, would arm the tanks for jettisoning or would jettison the tanks. I have forgotten exactly how the switch worked. I do recall that the pilot looked to see if it was in the correct position as part of the pre-engine start check. Otherwise, you left the switch alone.

Someone, could have been me, must have done something to the switch that shouldn't have been done. As I turned out of my parking spot, one of the tip tanks fell off without me knowing it. As I started my climb out after takeoff, the tower called and said, "F-84 (whatever my number was) are you aware that one of your tip tanks is missing?" I checked, and sure enough it was. Advising the tower that I knew it now, I returned and landed.

Actually, the F-84 flew fine with only one tip tank. If the tower had not called, I probably would have finished the flight and returned to the parking area without knowing that the tank wasn't there.

F-84

TEST PILOT SCHOOL CLASSMATE'S CLOSE CALL

Captain "Chuck" Klobossa, one of the seven fellow classmen, had an unforgettable experience. An engine problem forced him to belly land an F-86 in the rough terrain on the east edge of the Edwards Dry Lake. The crash jammed the F-86 canopy in the closed position; and the aircraft caught fire. Had a T-37 crew not seen his situation, Chuck would most likely have bought the farm. The T-37 pilot landed on a smooth part of the dry lake near the crash site. A cameraman from the T-37 ran to the burning F-86, bashed the canopy in with his camera, and pulled Chuck out. It wasn't Chuck's time to go.

13

Working for the Navy

SECOND HOLLOMAN TOUR OF DUTY

G raduation from Test Pilot School brought a reassignment to Holloman AFB. There were absolutely no complaints about returning to New Mexico. However, the project to which I was assigned made the return to Holloman a mixed blessing. Holloman would be my home base. But I would go directly from Edwards to China Lake Naval Ordnance Test Station for temporary duty with the Navy. Katie, Terry, Steve, and Tim would move to base housing at Holloman. As it turned out, the temporary duty lasted for about eighteen months. I wasn't away this long. From time to time, I flew to Holloman in one of the two F-100s that I had at China Lake. In addition, the Navy let me use drone F-6 Hellcats to go to and from Holloman quite a few times.

From a duty standpoint, the temporary duty with the Navy was about as good as it gets in peacetime. I had an eleven-man aircraft maintenance team and two F-100Ds assigned to me. I flew a Lockheed owned and maintained F-104A. I reported to Lt Col Wood at Wright Patterson AFB. He was the USAF Systems Project Office Manager for the Air Force Sidewinder Missile Project. The function of The Air Force contingent at China Lake was to provide a Sidewinder Air-to-Air Missile test capability, using Air Force high performance fighters. At the time, the Navy had nothing comparable to the F-104A. Col. Wood required a weekly activity and test results report from me. I finally persuaded him to let me give the report by recording. After a couple of the recordings, he expressed a preference for that reporting mode.

I slept and kept my personal belongings in the Visiting Bachelor Officers Quarters. Except for being awakened before 0500 hours on a regular basis by a misguided Marine Corps Unit doing close order drill in the dark, the Bachelor Officers Quarters were fine.

TESTING THE F-100/SIDEWINDER WEAPONS SYSTEM

Several months before I came on the scene at China Lake, the Navy had demonstrated that they had the best air-to-air missile available anywhere. By invitation, they had made a trip to Holloman to show General Davis and others what they could do. The Hughes Falcon had been in the development phase at Holloman for quite some time, with less than commendable results. So, the folks at Holloman were skeptical about any statements that a Sidewinder would eat up whatever jet target the Air Force had to offer. A Navy pilot, Glenn Tierney, was particularly outspoken about the Sidewinder performance. When Glenn backed his brag by shooting down a drone with the first missile fired, the Air Force didn't hesitate to dispatch a Holloman pilot, Lt. Roebson, to China Lake with an F-100.

By the time I replaced Lt. Roebson at China Lake, the Sidewinder testing focus was on high altitude and high speed launches. Two types of targets were available for tests above 50,000 ft, flares mounted on balloons and flares attached to five-inch, high-velocity aircraft rockets (HIVARs).

F-100/SIDEWINDER TESTS AT CHINA LAKE

The F-100D was powered by a Pratt & Whitney, J57-P-21A, engine. The engine delivered 11,700 lbs of thrust in military power and 16,500 lbs of thrust in afterburner. The combat ceiling was advertised to be 47,700 ft. This was an exaggeration if 20 mm cannons were being used. It was legitimate if Sidewinders were being launched. A "zoom" maneuver, trading kinetic energy for altitude, could be used to shoot down a target that was higher and faster as long as the range at the time of missile release was within the Sidewinder performance envelope.

F-100D

I probably launched more than one Sidewinder from an F-100 at a target rocket during the China Lake tour. But I don't recall doing it. I do recall knocking flares off of the wing tips of F6F drones with Sidewinders launched from an F-100. The Sidewinder guidance was so effective that the F6F usually returned home to fly another day. I never shot down an F6F. All but a very few of the Sidewinder tests using F-6 drone targets were done with telemetry units substituted for warheads.

By the Fall of 1957, flight tests had clearly demonstrated that the Sidewinder would take out low speed targets (F-6Fs), high subsonic speed targets (QF-80s), and supersonic targets (target rockets) at altitudes below 40,000 ft. At altitudes of 60,000 ft +, a target rocket was an easy target for a Sidewinder, when the missile was launched from dead astern.

Flares suspended from a balloon were another story. Most of the launches at balloons were made with balloons floating at about 50,000 ft. The launch profile was an after-burner acceleration to mach number around 1.2 at about 38,000 ft, followed by a "zoom" maneuver at an angle of around 45 deg. During the zoom, the pilot used his eyeballs and ears to decide if the flares were emitting sufficient infrared energy for the missile to acquire the target. A couple of problems made this a difficult task. First, the flares were not all that easy to see against the bright blue sky and the sunlight reflection from a shiny plastic balloon. Secondly, the sun's infrared energy, reflected from the balloon skin, was picked up by the sidewinder's seeker. Heat energy from the balloon skin sent an aural signal to the pilot's headset just as a hot target wood.

The pilot had about five seconds to decide whether or not the flares were burning. In other terms, each mission the pilot had one chance to be a hero. He had two chances to be a loser.

If he had a good target and launched, he was a champion.

or

If he had a good target and didn't launch, he was a bum.

or

If the flares were not lit and he launched, he was a loser.

Sometimes I was a champion. Other times I was a bum.

With rare exceptions, just about the time the pilot was hitting the Sidewinder launch button on the "zoom" missions, there was a wake-up call from the J-57 engine. At high angles of attack (near stall speeds) the J-57 would compressor-stall. To my knowledge, there was no likelihood that these damaged the engine. Neither did I ever experience a flameout when they occurred. But they surely did get the pilot's attention. I never straddled the barrel of a German 88 anti-aircraft gun when it was firing. But I'm convinced that the sound in your ears and the sensations on your rear end would have been similar to those delivered by the J-57 compressor stall. A roll-off into a nose down attitude brought the engine behavior back to normal.

THE REGULUS II KILL

In the '50s the Navy had Regulus I and Regulus II cruise missiles. Both were powered by jet engines. The Regulus I was subsonic, and the Regulus II was supersonic. Navy pilots flying F-9s generally did the Sidewinder launches when Regulus missiles were used as targets. Only once did I get an opportunity to join the "shooters". The occasion was a Sidewinder warhead fuse test, using a Regulus II. The test was conducted from Point Mugu.

The plan was to launch a Sidewinder from a trailing position just enough out of range to prevent the Sidewinder from hitting the target, but close enough to let the warhead fuse "see" the target and trigger the warhead firing mechanism. Two or three F-9s carrying Sidewinders were to be the launch aircraft. Since the F-9 was a Navy subsonic fighter, the launch F-9s had to be vectored into range by the ground controllers as the Regulus II came out of a turn and passed by the F-9s. Plan "B" was to have the F-100 circle in a holding position somewhere behind the F-9s. If for some reason, plan "A" did not work, the

F-100 would use afterburner to dash into position and make the Sidewinder launch. The Regulus was fuel limited. So, decisions had to be made quickly.

I took off from Oxnard AFB, nearby, because Point Mugu did not have a starter adapter which would fit the F-100. I don't recall how many circuits of the racetrack pattern were made by the Regulus II before the call came to send in the F-100. But it wasn't many. Once I had been vectored into position by Point Mugu Ground-Controllers, the critical decision to be made in the Point Mugu Control Room was whether the F-100 was a wee bit out of range or in range. The debate wasn't long, for a matter of seconds determined whether the test would be aborted or completed. The decision was made to chance it.

A few seconds after the Sidewinder left the launch rails, the Regulus II became a large fireball. Fortunately, at that time the F-100 was not equipped with a ranging-radar. The F-100 launch pilot was blameless. The F-9 pilots were very disappointed.

F-100 MAINTENANCE AT EDWARDS

The eleven-man ground crew that I had at China Lake was top notch. If they had the parts and the tools they needed, they could make the machine flyable. They were, however, sometimes limited by the tools they had in their toolboxes and whatever they could borrow from the Navy. For larger maintenance challenges, equipment and parts had to be flown in; or, if the F-100 was flyable, it could be taken to some place that did F-100 maintenance.

Edwards was, about fifty miles or so south of China Lake. So, the first time that one of the F-100s developed an engine problem, I landed at Edwards instead of China Lake. Since the aircraft was not in flying condition, Edwards Maintenance changed out the engine without much fuss.

Several weeks later, as I parked my F-100 on the China Lake flight line after a test mission, the crew chief gave me the engine shutdown signal with an air of "Let's get it shut down NOW!" As I climbed down the ladder from the cockpit, I could see a group checking out a pool of jet fuel on the ground near the aft end of the fuselage. It turned out that an afterburner pigtail had failed. Unlike the pigtail breaks experienced by Lt. Cohn and Buck Buchanan, the break on my F-100 did not produce an explosion. *Only one pair of footprints that day.*

Now we had a maintenance problem. The engine must be pulled to change out the pigtail. But the afterburner was needed for a takeoff. The altitude of the China Lake Airfield was about 3800 ft. The weather was hot. A check of the performance section of the F-100 Pilot's Handbook indicated that, if things went well, I could get the aircraft airborne with military-thrust (no afterburner) by using just about all of the runway. As a backup, if I discovered during the takeoff roll that I wasn't going to make it in military thrust, I would light the afterburner and hope that there was no explosion. When the big moment came, there was no afterburner light. But a block long dust trail was created across the desert while the landing gear retracted.

Fifty percent of the negotiations with the maintenance folks at Edwards (their half of the discussion) for a second repair job on the F-100 could only be described as "Downright Hostile". They finally did the work, after I took an oath that Edwards would not be the "airfield of choice" for any future maintenance.

DOUBLE TIRE CHANGE AT NELLIS AFB

A time came when the F-100 main gear tire supply dwindled to the pair that we had on the one F-100 in place at China Lake. Sidewinder testing with this F-100 continued until even the most cursory check of the main gear tires during preflight said, "Don't fly this aircraft." Having taken the oath at Edwards, I dared not try their supply for a pair of tires. So, I decided to make a flight to Nellis and get their maintenance troops to change out the tires.

There was, of course, a possibility that a tire would fail taking off, landing, or taxiing. It can be hot at Nellis even in the Spring and Fall. It can be very hot in the Summer. I was hoping for the best. But I never expected a happening bordering on a minor miracle. Everything went great until I made my last turn into a Nellis parking space and stopped. The instant the aircraft came to a halt, BOTH main gear tires blew out with a resounding bang. Call that a coincidence? Maybe one tire... but two tires?

BETWEEN A ROCK AND A HARD PLACE

Ever now and then, there would be a break in the requirement for the F-100 to do Sidewinder launches at China Lake. When this occurred, I usually flew an F-100 back to Holloman for maintenance or some other reason. Generally, I did this at the end of the China Lake workday. On one such occasion, I was flying an F-100 that had an inoperative attitude indicator. The weather prophet at China Lake advised that there was a front between China Lake and Holloman, with cloud tops going way up there. The front extended from the Mexican border to southern Colorado. So, there was no going around it. I decided to have a go at going over it. If things turned sour, I could always turn around and land at Luke AFB, near Phoenix.

I found myself on top of the front as I passed Phoenix and still climbing in military thrust to stay in the clear. In the distance it appeared that I could stay above the cloud tops. By now the sun was disappearing below the western horizon; and the edge of the earth's shadow was coming into a view ahead. No problem, as long as the stars are visible. The weather at Holloman was forecast to be clear.

About two thirds of the way to Holloman, with the throttle at military thrust setting, I found myself in the clouds. With no attitude indicator, I lit the afterburner, climbed to an altitude a couple of thousand feet above the cloud tops, and came out of afterburner. Within a couple of minutes, I'm back in the clouds. This cycle was repeated several times while a mental estimate of the fuel required to get to Holloman in this mode was made. The predicted fuel remaining over Holloman had a negative sign. Just as the "You jerk, you've flown up a box canyon." feeling was about to set in, I noticed that the time from afterburner light to cloud reentry was getting longer. Soon, there was no cloud reentry. Holloman was indeed clear when I made a routine letdown and landing.

SHARED INSTRUMENTS FLIGHT TO PLATTSBURG

China Lake had an infrared radiation-measuring device that could be mounted on an F-100. It was used to measure the infrared radiation signature of the engines on a variety of aircraft. The date is lost from memory. But once a Russian TU-4 was scheduled to come to the United States and land at some airfield on the northeast coast. The Navy folks at China Lake asked if I could fly to Plattsburg, N.Y., intercept the TU-4 as it came by, and get some infrared measurements. For some reason, I went by Holloman to join up with John Pitts in another F-100. During our first leg on the cross country to Plattsburg, we discovered that John's aircraft had an inoperative attitude indicator. My gyrocompass was not functioning.

We made our first leg VFR on top of broken clouds. The second leg to Plattsburg had to be over an overcast sky, which was solid from about 20,000 ft to 1000 ft. To make things more interesting, there was no alternate available that we could reach from over Plattsburg. We rationalized that the urgency of the mission justified extraordinary optimism about fuel consumption. This allowed us to specify an alternate on our instrument clearance. Realistically, we had about fifteen minutes to get on the ground after arriving over the initial fix at Plattsburg.

Plattsburg was a B-47 SAC Base. When we arrived over the initial fix at Plattsburg and entered a holding pattern, there were several B-47s in holding and approach patterns below us. Our announcement to approach control that our fuel state dictated that we begin the letdown and approach within ten minutes was as welcome as an incoming ICBM.

Nevertheless, they moved out the B-47s below us and cleared us for the approach.

Our relationship with the locals didn't improve when we advised that we would need the voice channel to ourselves during the letdown. Since John had no attitude indicator and I had no gyrocompass, we had to make a two ship formation let down and approach. John flew my wing and called out the compass readings so that I could make heading corrections.

We were greeted at the parking ramp by one very irate SAC colonel. He said a lot about our mental prowess, judgment, I.Q., and such. None of it was complimentary. I suspect that only my tale about a high priority mission involving the Russian TU-4 limited his corrective action to a verbal reprimand.

FLYING THE F-104 AT HOLLOMAN

In January of '57, I joined a class of four or five pilots at Palmdale to attend the Lockheed Company F-104 Ground School. Joe Walker, the well-known NASA test pilot who was later killed in the B-70/F-104 mid-air collision, was one of the class members. Later that month, I checked out in an Edwards F-104A.

The F-104A had a wingspan of 21 ft 9 in. The empty weight was 13,384 lbs. At a maximum takeoff weight of 25,840 lbs, the wing loading was 131.8 lb/ft². The "A" model was powered with a General Electric J79-GE-3A/3B engine. This engine delivered 9,600 lbs of thrust in military thrust setting. At maximum after burner setting, the engine delivered 14,800 lbs of thrust.

In March of '57, I began doing high altitude, supersonic sidewinder launches over the China Lake Test Range in a Lockheed Company F-104A, Number 736. Al Sexton, an outstanding gentleman and long time friend, was the Lockheed Project Manager. He did an exceptional job of directing the Lockheed ground crew that took care of Number 736 and supported the Sidewinder tests.

Number 736 was, in some ways, still in the development phase. The J-79 engine afterburner would not light above 25,000 ft. So, if the test mission required afterburner above 25,000 ft, the pilot had to go into afterburner below that altitude and leave the engine in afterburner until afterburner was no longer required. Also, 736 had no ventral fin at that time. The ventral fin took care of the marginal directional stability problem at Mach 2. A ventral fin was later added, when we took 736 to Holloman.

Lockheed XF-104 "Starfighter"
USAF Museum Photo Archives

The first test series with F-104A 736 was made up of launches at an altitude of 60,000 ft. The Sidewinders were launched from the F-104 right wing at flare equipped, target rockets launched from the F-104 left wing. I would acquire the target rocket shortly after it left its launch rail, evaluate the infrared radiation from the target rocket flares, and launch the Sidewinder from the other wing tip. At target rocket launch time, the F-104 was doing Mach 2 at 60,000 ft.

China Lake is close to one hundred miles north of Palmdale. This distance quickly became important. Since 736's afterburner would not light above 25,000 ft, the afterburner had to be left on from the start of the takeoff roll at Palmdale until the Sidewinder left the launch rail over China Lake. On the 60,000 ft and Mach 2 tests, as the Sidewinder left the rail, I immediately brought the throttle to idle and turned for Palmdale. The fuel remaining at launch time was 300 lbs. There was no fuel for a go around. A successful landing on the first attempt was mandatory.

About the second or third Sidewinder launch at of 60,000 ft and Mach 2, the F-104 engine flamed out as the Sidewinder left the rail. I was wearing one of the early pressure suits that rendered the pilot almost unable to bend an elbow when it inflated. Knowing that the suit would begin to inflate very soon, I rolled the aircraft over into steep dive to try to get below 50,000 ft before the cockpit lost pressurization. It worked. The suit didn't inflate. Now the question was, "Will the engine restart?" I deployed the ram air turbine (RAT) to provide hydraulic pressure if the engine didn't start and wound down. I then advised Ground Control to tell the tower to keep the runway clear in case I had to make a dead stick landing. There was no dead stick landing. The first restart try at about 35,000 ft worked. The landing was, of course, made at China Lake.

This happening caused Lockheed and others to reconsider the original decision to fly out of Palmdale. So, the Lockheed Support Team moved their equipment and No. 736 to China Lake. As it turned out, this was a wise move. The aircraft engine flamed out on all of the next three 60,000 ft, Mach 2 launches. On the plus side, I was able to avoid a pressure suit inflation; and the engine restarted at about 35,000 ft on all three missions. After the fourth flameout, we received a wire from Wright-Patterson directing that we quit flaming out the engine. The wire said nothing about stopping 60,000 ft, Mach 2 tests. We had to laugh a little. I don't recall what was done to the engine. But Lockheed (with a lot of General Electric help) did fix the problem.

The Vought F-8U people were not as fortunate. One of their pilots was firing Sidewinders from a company aircraft. They struggled with a similar flameout problem at subsonic speeds and low altitudes for a long time.

THE ONE THAT GOT AWAY!

Standard practice for flying F-6 drones at China Lake was to have two shoot-down airplanes airborne any time a drone was flying. There came a day on which we were scheduled to launch a couple of Sidewinders at an F-6; and no second shoot-down airplane was available. As always, all of the Sidewinder Project troops were keen on going. Citing the long list of F-6 drone flights that had been made with no need to shoot down a drone, those in favor of going persuaded the Operations Boss to go with one shoot-down aircraft. At launch time, the shoot-down aircraft moved away from the drone to a holding point where there was no possibility of being hit with a Sidewinder.

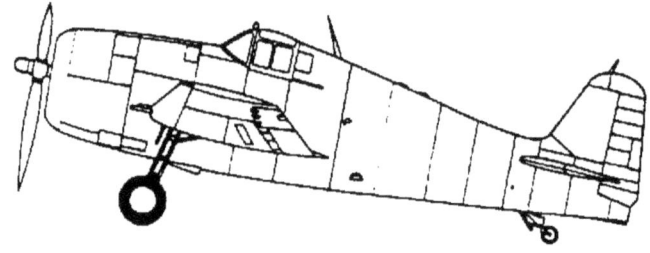

F6F Drone

The launches were subsonic. The launch altitude was somewhere around 8000 feet. The F-6 was fitted with wingtip flares and tail flares. I was to fly a fifteen or so mile wide racetrack pattern. The F-6 would be holding a straight south-to-north track up the range. I don't recall why the F-104 was being used. An F-100 could have flown the same pattern. After a dry run, I was vectored around for the firing pass. Usually, they used the wing tip flares first and the tail flares last. However, on this pass the tail flares were lit.

The Sidewinder scored a bullseye on the tail flare. There was nothing to indicate that the F-6 airframe had been touched. That is, the F-6 appeared to be in great shape until Ground Control tried to make a turn. There was no response to the turn command. By the time Drone Ground Control decided they needed help, both the shoot-down aircraft and the F-104 were not in position to do anything. The shoot-down aircraft had a considerable distance to travel to get into position for bringing the drone down with guns. I was halfway around the F-104 racetrack headed south, a good twenty miles from the drone location. I turned towards the drone location, put on the afterburner, and began an eyeball search for the drone. The shoot-down plane was also trying to catch the F-6.

As misfortune would have it, there was a front just north of the China Lake with a solid west-east line of clouds with tops several thousand feet above the drone altitude. Before either "shooter" could get into position, the drone disappeared into the cloudbank. Lots of communication links with ATC and others were very busy for several hours. A collision with a commercial airliner or a crash in the middle of Reno were just a couple of the terrifying possibilities. Finally, a report by a farmer in Washington State stopped the hand wringing. A red airplane with no pilot crashed on his farm. This time, when we lowered our contingency capability to get a test off, we got bit.

F-104 MAIN TIRE SHORTAGE

Even with Lockheed providing parts for No. 736, we found ourselves with a tire shortage. A Navy lineman chanced to hear us complaining about being grounded for lack of main gear tires. Looking over the F-104 main gear, he noted that the F-104 tires looked very much like the tires on one of the Navy aircraft on the flight line. It turned out that Navy Supply had quite a few spare tires for that machine. The Navy was always willing and ready to help us any way they could. In a very short time No. 736 was equipped with a brand-new set of tires.

Things are often not what they appear to be. That was true of our new set of wheels. The Navy tires were low-speed tires. The F-104 tires were high-speed tires. The 175 kt F-104 landing touchdown speed was outside the performance range of the Navy tires. We proved this to be the case by blowing tires during more than one landing. It turned out that directional control was not a problem with a blown tire. But the pieces of tire being thrown radially could damage the wing or other airplane parts. After a couple of blowouts, we quit using the Navy tires.

FLYING THE NAVY F-6 HELLCAT

A very enjoyable fringe benefit available during my temporary duty tour at China Lake was flying the F-6 drones. Navy rules required that an F-6 drone log eleven hours of autopilot time before it was flown without a pilot. Observing the autopilot fly an F-6 was not a duty for which the China Lake Navy pilots competed. So, I had no problem getting approval to fly the F-6 drones to and from Holloman on weekends. The autopilots had very precise altitude and compass heading hold capabilities. The autopilots held altitude within fifty feet of selected altitude. Compass heading was maintained within one degree.

HELLCAT

It was great to fly from China Lake to Holloman about five hundred feet above the terrain and take in the view of the wide-open spaces. A pencil sized stick on the right console allowed the pilot to fly the F-6 via the autopilot; or the autopilot could be allowed to just hold altitude and compass heading. Landing the F-6 with the pencil stick was a challenging exercise.

QUIET TIME

The route from China Lake to Holloman passed over the Gila Wilderness. I particularly enjoyed flying a circuitous route through a Gila Wilderness canyon that ran more-or-less east and west. The canyon had a beautiful stream running through it. Flying west-to-east, a climb out of the canyon had to be initiated far enough west to clear a north-south mountain range near Truth or Consequences.

The checkpoint for starting the climb was a ranch house and the other buildings that ranches have. Just to be friendly, I usually came by at tree top height and waved.

On one occasion, just as I pulled up after the wave, the engine abruptly quit. Silence at a couple of hundred feet above rocks, bushes, and trees is a real adrenaline generator. The old mind and body were anxious to do something to keep the F-6 airborne. But the airspeed and altitude said get the nose down. Don't hit the ground stalled! Suddenly, the engine started running again.

Leaving everything just as it was, I put the F-6 in a shallow climb and left it there until I was high enough to clear the mountain range by a wide margin. The engine ran normally the rest of the way to Holloman. Taking no chances, I made a dead stick landing pattern.

After I landed, I told transient maintenance about the incident, and asked them to please check the engine as best they could. The maintenance troops at Holloman had zero experience on a World War II Navy fighter.

When the time came to depart for China Lake, Holloman Maintenance reported that they could find nothing wrong with the aircraft. This wasn't very reassuring. But I was fresh out of ideas on where to get a second opinion. So, I fired up, taxied out, and took off. Just as the gear was locking into the wheel wells after takeoff, the engine quit. As I dumped the nose and prepared to put the airplane on the ground, I thought, "At least the terrain isn't as rugged as it was the first time. Once again, just about the time to start bringing the nose up for the flare, the engine started running. Cautiously, I flew a tight pattern and landed. This time, there could be no, "We couldn't find anything wrong." The maintenance troops had witnessed the entire happening. They finally located a loose ignition wire, fixed it, and sent me on my way back to China Lake.

F-6 INTEGRATED AIRSHOW

An Armed Forces Day came up on the calendar while I was TDY with the Navy. Someone came up with the idea of putting together an integrated three ship F-6 acrobatic team for Armed Forces Day. A Navy commander led the formation. A Marine Corp Captain flew left wing. I flew right wing. With only three or four flights to practice, we definitely were no threat to the Blue Angels or the Thunderbirds. Our one formation acrobatic maneuver was a loop. But we did it well. We thrilled the crowds at Mojave and at China Lake.

ASTRONAUT ASSIGNMENT OPPORTUNITY

During my stay at China Lake, a message was received which directed me to report to the Pentagon the next day. There was no clue in the message as to why. At the Pentagon, General White, the Air Force Chief of Staff at the time, briefed us about an opportunity to participate as flight crew members in NASA's planned manned space program. (The word, Astronaut, hadn't been coined yet.) The group was made up of Air Force test pilots, most of whom I knew.

After General White's briefing, we joined Navy and Marine Corp test pilots in a red brick building somewhere in Washington. Glenn and Shepard are the only Navy test pilots I remember meeting. For the next two days, NASA briefed, tested, and interviewed us. I recall asking in a Question-and-Answer session with the NASA troops, "What do we do about flying while we are with you?" The answer was, "We think we can get you excused from flying." The test pilot audience broke up laughing.

Comparing my outstanding setup at China Lake with what I had heard from NASA, I was on the fence about what choice to make on departure day. Buck Buchanan and I had a brief discussion about the pros and cons of the NASA opportunity while waiting for our departure briefings.

NASA interviewers and testers had raised their eyebrows several times about my deaf ear. I told Buck that I was considering giving it a "Go." However, if they said anything further about my deaf ear, I would tell them to forget it. The first statement by the NASA exit briefer was, "Your qualifications are excellent. We are a bit concerned about your deaf ear." My response was, "Forget it."

At the time, my vision of where manned spacecraft flight could lead was almost nonexistent. Had manned space operations been limited to orbiting the earth in a capsule, I would have no regrets about not competing. Considering NASA's unbelievable accomplishment of sending men to the moon and current Space Shuttle Operations, I passed up a great opportunity.

14

Sidewinder Testing

RETURN TO HOLLOMAN - 1957

B y the Summer of 1957, flight tests of the Sidewinder using F-6 drones, QF-80 drones, and Regulus Cruise Missiles at altitudes from 1,000 ft to 40,000 ft had clearly demonstrated that it was the best air-to-air weapon in anyone's arsenal, worldwide. If one accepted target rockets as legitimate targets, the Sidewinder had demonstrated a high kill ratio at altitudes up to 60,000 ft. There were, however, skeptics who did not agree that target rockets were a sufficient challenge. It was true that all the launches made at target rockets were from dead astern. And it was true that the target rocket started a gravity descent as soon as they came off of the launch rail.

Balloons with flares were still the only alternative targets available at altitudes above 40,000 ft. And our success rate with these was still zero.

Q-2 drones were brought from Holloman on the wings of the B-26 launch aircraft. The Q-2 was designed to recover itself by parachute if radar contact with the onboard transponder was interrupted for more than ten seconds. The first try with the Q-2 ended in the drone parachuting into some very steep mountains east of China Lake. Capt Jonesy Seigler, flying a helicopter from Edwards, recovered it from the side of the mountain.

The ground slope where the Q-2 was located was about 45 deg. A ground team from China Lake installed a cable and hook on the Q-2. I positioned myself such that I could reach the connector on the bottom of the helicopter. Jonesy positioned the helicopter and held it in a hover with the rotor blade missing the ground by about ten feet while I snapped the hook in place. We had asked Edwards for their best helicopter pilot. That's who they sent.

About this time, it became apparent that the need for the Air Force F-100, F-104 team at China Lake had just about run its course. China Lake had committed to helping establish a Sidewinder test capability at Holloman. So, it was time to go home.

SIDEWINDER TESTING AT HOLLOMAN

The Navy outdid themselves by helping the Air Force Missile Development center establish a Sidewinder test capability at Holloman. They essentially gave the Air Force a complete Sidewinder assembly, maintenance, and checkout facility. They also trained some very bright enlisted missile experts that we already had at Holloman on how to operate the facility. The training included assembly, maintenance, and checkout of the Sidewinder telemetry system. To be sure that we came up to speed quickly, the Navy also assigned an experienced Philco Tech Rep, Gerald Strome, to the Holloman Sidewinder Project.

I was named Chief of the Holloman Sidewinder Project, reporting to a great boss, Major Robin Hansen. One time when things became stressful, he fired me one day and hired me back the next.

Running the Sidewinder Project involved planning the test program, scheduling the tests, flying tests, and getting test reports produced. A Lt. Ed Bauman was assigned to the Sidewinder project. Ed did a super job as a one-man flight test engineering and data analysis office. He was analytically sharp and very goal oriented. Major John Pitts was Chief of the Fighter Missile Test Branch. This was the outfit which provided the missile, telemetry, and aircraft launch system support for getting a Sidewinder test accomplished. Captain Ed Moore was John's right-hand man, and a fine test pilot. John and Ed flew some of the Sidewinder tests. Aircraft operations support was provided by a superb Aircraft Maintenance Crew from the 6580th Field Maintenance Squadron.

Once the Sidewinder Lab became operational, Air Force Sidewinder testing began at Holloman. F-104A No. 736, F-100 No. 144, and F-100 No. 138 were moved to Holloman. We also acquired another F-104C launch aircraft, No. 757. For a few months after the move, we supported on-call Sidewinder testing at China Lake. When the Navy acquired their own F-104, the on-call operations at China Lake ceased.

MORE SIDEWINDER LAUNCHES AT FLARE CARRYING BALLOONS

The Navy was still very much interested in the Air Force test program, because we put a fair amount of emphasis on testing the Sidewinder at altitudes above 50,000 ft and at high launch Mach numbers. In fact, China Lake provided modified Sidewinders for us to test in the high altitude, high Mach number part of the operational envelope.

Initially, all of our launches at targets above 50,000 ft were made at flare-carrying balloons. Other than target rockets, balloons fitted with flares were the only high-altitude targets available to us. There was no reason to shoot more missiles at target rockets. Plenty of data was in hand which conclusively showed that target rockets fired from the Sidewinder launch aircraft were a piece of cake. We were fortunate that there was an outstanding balloon launch outfit already in place at Holloman. On the average, they could put a balloon inside a five-mile diameter circle at 60,000 ft or higher within plus or minus thirty minutes of a scheduled launch time.

John Pitts, Ed Moore, and I launched many Sidewinders at balloons from F-100s in zoom maneuvers and from the F-104s at the same altitude as the target balloon. We never came within Sidewinder fuse triggering range even once. The Navy changed guidance gains, control fin sizes, and other fixes. The only change was that occasionally a missile went unstable.

IN-HOUSE HIGH-ALTITUDE TARGET DEVELOPMENT

With no high-altitude targets other than balloons and target rockets on the horizon, I finally decided that we should try building a target ourselves. Ed Bauman was always looking for a new challenge, so we designed a plywood delta target powered with 2" HIVAR motors. The target had two delta aerodynamic surfaces. The horizontal delta surface was the wing. It had "settable" trailing edges that could be fixed for level flight or climb. The vertical delta surface provided directional stability.

The plan was to launch the target from a balloon or from an F-100 in a zoom maneuver. This meant that we also had to design and fabricate a launch rail for the balloon and a launch rail for the F-100.

Ed Bauman designed a system for the balloon launch-platform, which would keep the platform always pointing north. Two photoelectric cells with about a 5-degree dead band between them provided a beam of light on each side of the north heading of a compass mounted on the launch platform. If one of the beams of light was interrupted by the compass needle, a signal was sent to a geared electric motor which drove the launch platform back to a north heading.

Both of the target projects were un-funded. So, we had to scrounge the materials wherever we could find them. The compass we used was from a B-26. Since the project was not official, we did our ground testing after duty hours on a piece of ground north of King One.

Our test facility was comprised of two lengths of a small telephone pole set in the ground. Wire suspended the launcher between the poles as it would be attached to the balloon. This allowed us to test the system for holding the launcher on a north heading, to test the separation of the target from the launcher, and to verify our setting of the trailing edge flaps for the desired flight profile (level or climbing trajectory). For the first test, we fixed the trailing edge flaps at a trailing edge up position for a shallow climb.

About an hour after the range closed for the day, five or so of us mounted the target in our test facility, loaded the target on the launcher, strung about thirty feet of wire from the rocket motor igniters to the ignition button, and hunkered down behind a makeshift shield. Once the target left the rail, we planned to stand up and watch it follow a climbing trajectory up range.

The firing went great. The departure from the launch rail came off without a hitch. As planned, the target headed due north. That was the good news. The bad news was that the climbing trajectory was not at all what we expected. With our mouths hanging open in amazement, we watched the target execute the first half of a large diameter loop. Then, as the target began the second half of the loop, we realized that it was returning to the launch site. Almost at the same instant, everyone started scattering like quail. It quickly became apparent to all that we were wasting our energy. The target speed was far too great for us to improve our odds by running. So, we came to a halt and watched. The target impacted about a block south of the launch site.

We changed the trailing edge flap setting for the second launch.

FIRST BALLOON LAUNCHED DELTA TARGET TEST

The day came for the first Sidewinder firing against a balloon launched delta target. The first test was intentionally made simple. The target would make a shallow climb to the north. The F-104 launch aircraft would be approaching from the south at the same altitude as the balloon. On later flights, the F-104 would pick up and track the launched target in an angle off approach.

The Controller at King One vectored me into position to start the inbound track. With the Controller's help, I visually acquired the target. When the flares on the dart target were ignited, the solid audio sound in my earphones confirmed a good infrared signal from the Sidewinder seeker. Rechecking that the missile arming switches were "HOT", I held my control stick finger lightly on the firing button. The Controller started the countdown for an in-range target launch, while I waited for the Delta Target to separate from the balloon.

The plan was to track the Delta Target for just a few seconds after it cleared the launcher and release the Sidewinder. The Sidewinder launch never occurred. The instant the Controller at King One said "ZERO" for target-motor firing, I had a Delta target coming straight at me. The rolls of the target and the attacker had been reversed. My reflex action was a jerk on the control stick to change the F-104 flight path, not a push on the Sidewinder launch button. On later launches we did better.

MATING THE DELTA TARGET WITH THE F-100

Master Sergeant Tixer, the NCO to whom Major John Pitts assigned the responsibility for running the day-to-day activities of his Fighter Missile Test Branch, personally took charge of getting a Delta Target launcher for the F-100 built. In my opinion, he did a super job. We were within spitting distance of getting a test mission airborne, when one of the Center's roving inspectors happened to see the launcher on the wing of one of our F-100s and blew the whistle. Since it was an un-funded, unapproved (not disapproved, just unapproved) project, the launcher lacked a proper pedigree. Getting a pedigree would have taken longer than the Sidewinder Project lifetime. So, we made a tactical retreat and began a search for another alternative.

Quite unexpectedly, the alternative was handed to us. Colonel Wood, our Systems Project Officer (SPO) at Wright- Patterson AFB, called and told me to meet him at the Curtis Wright Company, located in Santa Barbara, California. He had found a modest amount of money for purchasing high altitude targets if some could be found. Someone had told him about a Dr. Vogt who worked for the Curtis Wright Company and who had a reputation for inventing and designing unusual aeronautical vehicles. He had been one of Hitler's design experts during WWII.

We spent a day at the Curtis Wright Company, about an hour of it with Dr. Vogt. He was an interesting little guy, and obviously very bright. He couldn't have been more than five feet tall. We spelled out our requirements to him in terms of flight altitude (60,000 ft minimum), Mach number (at least Mach 1+), Flight duration (two minutes), launch method (from F-100 wing), cost (cheap), and delivery date (a couple of months).

Most aircraft companies would have laughed us out the door. He listened attentively and then thought about what we said for a few minutes before responding. All he said was, "Ja, I think I can provide you what you're looking for. Come back in about four weeks." When we went back, he had an Army anti-tank rocket with stabilizing fins on the back and a settable canard just aft of the rocket nose. The stabilizing fins had small-chord, movable flaps that controlled roll. The flaps oscillated continuously with very small amplitude. This avoided any undesirable step roll commands due to break out friction. The canard was locked in place before the target was loaded on the aircraft. The canard setting angle determined whether the target flew level or climbed after leaving the F-100 launch rail. The Curtis Wright Company delivered something like fifteen of them over a period of two months. We didn't have a single failure.

AT LAST!! SUCCESS AT HIGH ALTITUDES

When launched from an F-100 in a zoom maneuver, Dr. Vogt's target did everything that he promised. It flew level or it climbed depending on the canard setting we used, making possible angle off Sidewinder launches at whatever altitudes we could fly the F-104. At altitudes up to 60,000 ft, when launched inside a cone of about thirty degrees aft of Dr. Vogt's target, the sidewinder was very effective. We conceded that the Sidewinder couldn't handle zero velocity targets at those altitudes and terminated our firings at balloons.

HIGH CLOSURE RATE SURPRISE AT 30,000 ft

In parallel with high altitude testing, launches were made to further define the Sidewinder launch envelope at altitudes from sea level to forty thousand feet. One high closure rate launch at a QF-80 from dead astern (zero angle off) turned out to be more than was expected.

The plan was to fly the QF-80 due north at about 25,000 feet and Mach 0.5 The F-104 launch aircraft would close at Mach 0.9 to the maximum Sidewinder range position and launch the missile. As soon as the missile left the launch rail, a break right or left would be executed to avoid the QF-80. Everything went as planned until the F-104 reached the point where I was to visually acquire the target. With all the help the Ground Controller could give me, I just was not seeing it. Early visual acquisition was particularly important, because of the high closure rate. Just as the Ground Controller called for an abort, I saw the QF-80 very close and directly ahead and hit the Sidewinder release button. Before the collision avoidance maneuver could even be started, the QF-80 turned into a huge ball of fire. The F-104 went straight through it.

Amazingly, there were only three places that pieces of the QF-80 hit the F-104. None of the damage was serious. In addition, the F-104 engine behaved as if nothing had happened. Carried again!

COUNTERED COUNTERMEASURES

In late 1958, Sidewinder's successes captured the attention of the countermeasures world. At Holloman, this brought into being a project dedicated to breaking the Sidewinder lock on the tailpipe of a QF-80 and leading it elsewhere. The plan was to eject flares from the QF-80 as the Sidewinder was on its way for a rendezvous with the tailpipe. The infrared radiation from the flares was to be substantially higher than the tailpipe radiation. Consequently, the flare's gravity free fall flight path should decoy the Sidewinder away from the QF-80.

As pilot of the F-104 launch aircraft and a strong advocate of having Sidewinders on every Air Force day fighter, some countermeasures folks suggested that I biased the test results by timing my Sidewinder releases to render the flares ineffective. I was flattered by their suggestion that I could do this. However, all I could say was that when the Sidewinders came off the launch rails, I was looking at the QF-80 tailpipe and the flares through my sight ring.

The countermeasures project was allocated two QF-80s. The project lasted for two Sidewinder launches.

ANOTHER INTERESTING HIGH-ALTITUDE TARGET

While Sidewinders were being tested using F-100s and F-104s, a much larger unguided missile was being tested, using an F-102. That project also needed a high-altitude target. Their target requirements were much more demanding than ours. The target had to fly at altitudes above 50,000 ft long enough for the launching F-102 to acquire the target with radar at a range of fifteen to twenty miles and track it to the missile launch point. What they needed was a drone aircraft at high altitudes. Unfortunately, they had neither the time nor money to develop and acquire one.

I noted earlier that General Davis encouraged creative thinking. He always listened with an open mind to ideas that others would have immediately tagged as too radical. It was, therefore, not surprising that two engineers in the aircraft Missile Test Directorate were given the go ahead to examine the risk associated with using manned F-104s as targets.

The two engineers did a superb engineering analysis and sales job. So that non-engineers would better understand their analysis results, they showed that the risk to the F-104 and pilot was less than the risk associated with an employee driving his or her automobile from the Holloman AFB main gate to downtown Alamogordo at quitting time. This was a distance of six miles on a straight road. The plan was approved.

I was among several F-104 pilots who flew the target F-104s. To my knowledge, no one ever had a close call. It was interesting to watch it all happen from the target F-104 cockpit. The intercept geometry was to have the F-104 crossing the F-102 flight path at an angle. About ten miles from the crossing of the two flight paths, the F-102 fire control radar would lock on the F-104 and provide the necessary F-102 guidance to the missile release point. As long as the missile was still on the F-102 launch rail, an audio signal would be transmitted to the F-104 pilot's earphones. When the missile left the launch rail, the tone would immediately stop. The instant the audio signal ceased, the F-104 pilot pulled up sharply. By rolling over a little, you could watch the missile go underneath you.

Ground tracking Askanias were recording both the F-104 and the missiles' trajectories. It was a simple data reduction exercise to determine what the missile's miss distance would have been if the F-104 had not pulled up.

SORE BACK FROM AN F-104B HARD LANDING

Maintenance test flights were always fun things to do. When an aircraft came out of maintenance, it was taken up for thorough systems checks before it was placed on the flight line for operational flying. Generally, fuel remained after all of the required systems checks were completed. There was time to do whatever tickled your fancy.

One day, as I was nearing completion of the required systems checks on a F-104B (two-seater F-104), I went into afterburner at an altitude of about 35,000 ft to accelerate to maximum Mach number. Just as the speed went past Mach 1, the control stick became rigid. I immediately came out of afterburner, and applied all the muscle I had in an attempt to move the stick. It didn't move a millimeter. Just as the mind went into the "What now?" thinking mode, the stick suddenly was normal again. Before I could feel better about things, the stick returned to a rigid state. This cyclic stick behavior then became the steady state operation.

By this time I had headed for Holloman, and had advised the control tower of my problem. The tower, of course, did all of the proper things. The fire trucks and rescue folks deployed. F-104 systems experts came up on the communication link. And the "What do we do now?" got serious. Whatever was to be done had to get underway soon, for my fuel state was rapidly becoming critical. Bailout was an option as long as the option was exercised above an altitude of 1500 ft. The early F-104s had downward ejection.

Since pitch control was coming and going, I decided to try landing. Lowering the landing gear would result in a trim change. So, I decided to put the gear down at altitude. Things became more complicated when the gear handle was placed in the down position. The F-104B gear, unlike the F-104A gear, extended forward. Wind force on the nose gear would not let the gear free fall into the locked position. There was no hydraulic pressure to extend the nose gear. By now the fuel remaining was getting seriously low. There was no hesitation when the thought of snapping the gear into place with the landing drag chute entered my mind. Putting the nose of the F-104B well down, I yanked the deployment handle. After the nose gear snapped into place, I jettisoned the drag chute.

With the stick freezing and unfreezing, I managed to get the aircraft on a long straight in final at about a thousand feet. With three hundred pounds of fuel remaining and no bailout option, some sort of landing was going to occur. About halfway down the final, the stick froze and stayed that way. Still, things were going well. The aircraft was trimmed in a stabilized descent of about five hundred feet per minute. If I hadn't made a dumb mistake at the last second, I could have been a hero.

Habit did me in. An F-104 was one of the easiest airplanes to land that I have flown. It landed hot; and the flare was started at about three hundred feet. However, the ground effect at touchdown was like a giant cushion. The tires squeaked on just about every landing. My practice had been to come in a few knots high, chop the throttle in the flare, and bring the sink rate to zero. As the F-104 came over the end of the runway, I inwardly smiled and said to myself, "You made it!" I also chopped the throttle. Immediately, the nose dropped sharply.

There was no pitch control available to compensate for the trim change. The F-104 hit pitot boom first, bounced upward until the fuselage was almost perpendicular to the runway, came down on the tail, and slammed forward on the runway. With the gear spread flat, the remains slid down the runway several hundred feet and came to rest.

The little bit of fuel that remained in the ruptured tip tanks caught fire. The fire created no real hazard. It did, however, motivate me to get out of the cockpit in hurry. I was, in fact, in such a hurry that I forgot that my flying boots were still attached to the cables which pull the pilots' legs aft for a downward ejection. Before exiting the cockpit, the pilot must rotate his heels about his toes to disconnect the cables. Opening the canopy and turning my body sideways, I made a leap for the outside. I wound up hanging head down outside the cockpit still attached to the cables. Getting back in so that I could detach the cables was probably the most difficult part of the entire happening.

Except for a very sore back for a couple of weeks, I suffered no lasting damage. The F-104B was totaled. An improperly connected hydraulic line to the slab horizontal tail had dumped the flight control system hydraulic fluid.

COMPRESSOR STALLS AROUND THE TRAFFIC PAT-TERN

Engine compressor stalls in an F-100 never become routine events. But, after experiencing several, they don't come as a surprise when one is making a zoom maneuver at 50,000 feet. Engine compressor stalls in the traffic pattern are something else.

Once, when on duty in Mission Control at King One, a call came in that a chase aircraft was needed immediately. I was the most available pilot around. After advising them to have my parachute and helmet at the F-100, I mounted my scooter and, at maximum scooter speed, went directly to the chase aircraft. Running around the F-100 to make sure that all of the landing gear pins, etc. had been removed, I told the Crew Chief to start the APU, donned my flying gear, and strapped in.

The engine start, taxi, and takeoff roll were uneventful. In fact, everything went normal until the gear started up into the wheel wells. Suddenly, there was a king size compressor stall. The first one was immediately followed by another. The ground was too close to drop the nose. I did come out of afterburner and retard the throttle as much as I thought was prudent. The compressor stalling stopped. Each time I would try to advance the throttle to military power setting, the engine would compressor stall. Fortunately, there was enough thrust to get the aircraft around the traffic pattern at a throttle setting just below that which produced compressor stalls.

As I taxied back to the flight line, I advised the tower to tell the flight line personnel to have another F-100 ready. There was no problem with the second F-100. The test mission was completed.

The crew chief of the first F-100 was waiting for me when I taxied in. He was holding the large wooden board that is placed about eight inches inside the F-100 engine air intake to keep out dust and debris when the aircraft is parked on the ramp. A pilot of average height must chin himself on the lower lip of the air intake to see it during preflight. I had failed to check for a nose plug during my run around before climbing into the cockpit.

During engine start, taxi, and takeoff, the wooden board had turned sideways and allowed normal engine operation. About the time the gear handle was raised after liftoff, the board came loose and slammed flat against the row of stator blades at the front of the engine compressor. This substantially reduced the airflow through the engine and produced a compressor stall. Carried again!

The incident did have a good side. Throughout the Air Force, wooden nose plugs were replaced with canvas nose covers that could be seen easily from anywhere in front of the F-100.

15

Aerospace Research Pilot School

EDWARDS AFB - SECOND TIME

In the Fall of 1959, everyone who had a say decided that the Sidewinder missile testing was complete. Holloman was still a busy place with interesting things going on; but there were no jobs and/or flying opportunities that compared with what I had been doing for a couple of years. What does one do when there are no good opportunities? He goes to school. So, I applied for and was given a slot at the University of Michigan to get a PHD in engineering.

The week came for the moving vans to pick up our worldly possessions and haul them to Ann Arbor. That same week, Major Dick Lathrop, the USAF Test Pilot School Commandant, dropped by to offer me the Operations Officer Job at the Test Pilot School. After a sleepless night of mental gymnastics, I called the Director of Personnel at Air Force Systems Command Headquarters, and requested some time with him to discuss where I should be assigned. The next day I flew a T-33 to Andrews AFB to present my case.

The Director of Personnel was not at all in favor of changing my assignment to Edwards. So, his agreement to make the change came as a surprise.

The six McElmurrys packed up and moved to Edwards in December 1959. Terry, Steve, and Tim had lived there before. This was Mike's first move. Our first house was an improvement over the house we lived in the first time we were stationed at Edwards. A few months later, we moved across the street to an even better place.

The former tenant in the second house had a green thumb. He left us a grape vineyard with several kinds of grapes and a strawberry patch.

Being Operations Officer of the Test Pilot School was great duty. Dick Lathrop did a first-rate job as Commandant. It was a pleasure to work for him. Every member of the staff was enjoyable to work with. They were all capable, self-starting performers.

Martin RB-57A 'Canberra'
©USAF Museum Photo Archives

The Test Pilot School aircraft inventory still included T-28s, T-33s, F-86s, and an F-100C. The B-25s and the F-84 were gone. They were replaced with B-57s, F-104Bs, and an F-102A. I had never flown the B-57 or the F-102A. So, it was great to get time in them. The B-57 engines were started with cartridges about the size of a water bucket. They created a huge cloud of black smoke when they were ignited. With a full fuel load and a good tail wind, a flight from Edwards AFB to Andrews AFB (west coast to east coast) could be made without an intermediate stop for fuel.

The F-102A had good handling qualities and no significant undesirable characteristics. There was no buffeting or tendency to have a wing drop in a stall. Instead, at stall angles of attack with the stick held in the full aft position, the

F-102 would descend at 6000+ ft/min. Some care was advisable on landing finals to avoid inadvertently getting into a high sink rate and bending the landing gear. I once watched an experienced company test pilot do this in a landing at Holloman AFB.

CLOSE CALL IN AN F-102

Carelessness almost cost me my life during an F-102A takeoff at Edwards. The frequency selector for the tacan in our F-102A was located near the floor on the left side of the pilot's seat. I don't recall why there was a need for the tacan immediately after takeoff. But as I raised the gear, I focused my eyeballs on the tacan channel selector near the floor and began trying to set it. Very few seconds could have elapsed before an inaudible voice in my head said, "Look up!" When I did, there was just enough time to recover from a nose down, ninety-degree bank and miss the lakebed surface by 20 ft or less. The Accident Investigation Board would never have been able to figure out what happened.

F-102

FLYING THE HUEY HELICOPTER

Class 60-B at the Test Pilot School was a class made up entirely of Army helicopter pilots. Class 60-B is the only all Army class that has been conducted at the USAF Test Pilot School. To provide the helicopter experience needed for such a course, Major Jonsey Seigler, the helicopter pilot who lifted our downed Q-2 off of the mountainside near China Lake, replaced me as Operations Officer. Jonsey gave me a ten-hour course in a Huey Helicopter. He was an excellent instructor. By the end of the ten hours, I could do an auto rotation to touchdown, hover, take off and land. My first attempts at doing any of these maneuvers were humbling experiences. Jonsey laughed a lot, watching a cocky fighter pilot get his ego knocked down several notches.

THE AEROSPACE RESEARCH PILOT SCHOOL

When Jonsey replaced me as Operations Officer, Dick Lathrop gave me an assignment as Special Assistant to the Commandant for structuring and conducting a course to train test pilots for spacecraft testing. Dick provided quality help but almost no resources to accomplish the task. The quality help came in the form of Mr. Bill Sweickhard, an exceptional flight test instructor and flight test engineer. The resources consisted of a T-33 for unlimited cross-country flying, no funds, and any free help we could get from Aerospace Companies and/or Government Agencies.

"No funds" included no reimbursement for lodging, food, or other expenses incurred on cross-country trips. Considering the constraints, we fared pretty well.

- Chance-Vought in Dallas agreed to give us free time on their moving-base simulator.

- North American Aviation in Los Angeles donated time on a fixed base simulator.

- The zero "g" aircraft folks at Wright-Patterson AFB blocked out some free time in their zero "g" aircraft for us.

- Boeing Aircraft in Seattle scheduled a free show-and-tell briefing on their Dynasoar Program.

- We were given free participation in a Space Seminar at the University of Michigan.

- We were loaned a set of films which covered a Space Seminar conducted by Von Braun and others of his stature.

- Ames Research Center used Borman, McDivitt, and I as subjects in a Mercury Spacecraft reentry control task performed while experiencing various "g" forces generated by the Navy Centrifuge at Johnsville, Pennsylvania.

I am sure that there were many other worthwhile contributions to the course. But all that was thirty-seven years ago. The memory trail back that far is in many places covered with a lot of overgrowth.

THE FIRST AEROSPACE RESEARCH PILOT CLASS

Graduates of the Test Pilot School were the best available candidates for a course designed to teach space vehicle testing. Frank Borman had just graduated from Test Pilot School with an excellent record. He agreed to be a member of the first Space Research Pilot Class. Dick Lathrop asked who we should get from the Edwards Flight Test group. I told him that Jim McDivitt would be my choice. Jim was not at all keen on giving up a good assignment in Fighter Test to return to the School as a student. However, when he was drafted, he came with his customary positive outlook and "gung-ho" attitude.

Buck Buchanan was about to finish his work at the University of Michigan for his PHD Degree. The plan was to have him join the class late as the fifth student/instructor. Buck almost didn't make it. In May 1961, I received an urgent telephone call from Buck, advising that he had received orders to Maxwell Field when he finished at the University of Michigan. I flew a B-57 to Selfridge Field, picked him up, and flew to Andrews Air Force Base. Fortunately, the Air Force Systems Command Personnel Chief was not the same colonel that I had talked into letting me out of my assignment to the University of Michigan. We were able to persuade the new guy that it was absolutely essential that Buck be reassigned to Edwards AFB.

Office of Information
Air Force Flight Test Center
Air Force Systems Command
Edwards Air Force Base, Calif

Major Thomas D. McElmurry, Chief, Aerospace Research Pilot Div., Aerospace Research Pilot School, Air Force Flight Test Center, Edwards AFB, Calif., is one of five men graduating December 15, 1961, from the first Aerospace Research Pilot Course. His home is Batesville, Arkansas.

—U.S. Air Force—Aerospace Power for Peace—

CORNERING THE MARKET ON F-104s

Clearly, the F-104 was the ideal aircraft for simulating the approach and landing part of a spacecraft's flight profile. Once the Aerospace Research Pilot course had been approved by General Schriever, the boss of the Air Force Systems Command, there was no problem requisitioning F-104s from outfits all across the country. We had a flight line full of them for the first and subsequent classes.

F-104 ENERGY MANAGEMENT MANEUVERS

Managing the excess energy available in an F-104 at Mach 2 and an altitude of 38,000 ft to execute a planned maneuver is not in the same league as using the energy available from a spacecraft booster to place a spacecraft vehicle into orbit. Still, given a starting Mach number and altitude condition, there is an optimum F-104 flight path that can be calculated and flown to achieve a specific flight condition. As one of the flight training exercises for the Aerospace Research Pilot Course, we optimized zoom maneuvers to reach the maximum possible altitude. The zoom starting altitude was varied some. The zoom starting Mach number was always Mach 2, because that was the maximum F-104 Mach number allowed. The zoom-angle was selected by the pilot planning the flight. As I recall, it was about sixty degrees.

As the F-104 passed through 70,000 ft, the engine was shut down to avoid over-temping it. The nose was allowed to fall through going over the top. On the way down, the engine was restarted at about 35,000 ft. The flight path was flown such that, if the engine didn't restart, the aircraft was in position for a lakebed dead stick landing.

FRANK BORMAN'S DEAD STICK LANDING

Frank Borman was the only one in the class who had to make a dead stick landing. This occurred before the zoom maneuver was initiated. As he was accelerating to Mach 2 to start the maneuver, his fire warning light came on. Retarding the throttle didn't extinguish the light. So, doing what the Pilot's Handbook advises, Frank shut the engine off. Checking altitude, speed, and distance to the lakebed, he concluded that his energy state would put the aircraft on the ground short of a smooth part of the lakebed. So, he restarted the engine. Immediately the fire warning light brought a distinctive red glow to the instrument panel. Engaging in a bit of Russian Roulette, Frank let the engine run until there was no doubt that he could reach a suitable landing surface, and then shut it down. Fortunately, the fire that burned a king-sized hole in the aft fuselage didn't cut any flight control hydraulic lines. The dead-stick landing turned out great!

McDIVITT DRAWS THE BLACK BEAN

Just about every military pilot has at one time or another reported an abnormal behavior in the air that could not be duplicated by the maintenance troops after the flight. All of the four pilots in Class 1 of the Aerospace Research Pilot School experienced and reported an in-flight anomaly that behaved like this on one of the School's F-104s. In fact, this happening was reported more than four times. The entry in the Maintenance Record was always the same, "Flight control system functions normally during ground tests." Finally, we pilots just decided to live with the problem and quit writing it up.

In flight and sometimes during taxi, the stick would jerk to the left or right about half an inch. In flight, the F-104 would respond with a small "step" roll in the direction the stick moved. The happening was entirely random. Sometimes it happened on takeoff, sometimes during cruise, or on the final turn during landing. The frequency of the occurrence was also random.

Just about the time we all became "conditioned" to the little stick jerk and aircraft roll behavior, McDivitt was confronted with a serious challenge during his turn to final for a landing. The stick didn't just give a little kick and small roll response. It gave large jerks both left and right continually. Jim gave it his best effort to flare and touch down on the runway, right side up. He managed to get it flared with his head pointing skyward. But touchdown was in rough terrain. The F-104 fared poorly. Jim came through in good shape. The postmortem autopsy of the airplane revealed that the lateral control hydraulic system was messed up with contamination.

THE ROCKET AUGMENTED F-104s

About halfway through the course, Bill Schweikhard came into the classroom that we had converted into office space with the news that Base Salvage had six surplus J-2 rocket engines that were up for grabs. I told him to get the required paperwork together, pick them up, and bring them to our maintenance hanger.

Earlier, when we were at Wright Patterson for the "free training" we had obtained there, we had spent an entire evening in the motel where we were staying defining the facilities and equipment we would request from Systems Command and USAF Headquarters. We covered most everything in the simulation and aircraft categories. With no knowledge of rocket engine availability, the possibility of a rocket-powered aircraft never came up.

Now that we had the engines, putting some in F-104s was the obvious thing to do.

We still had no money. And, we were much too busy with the course to produce a conceptual design, make cost estimates, etc.

That needed to be done if we were to get a project approved by Systems Command and USAF Headquarters. We did the logical thing. We asked Lockheed, Burbank, if they would produce the things we needed for free. The smell of money is a powerful stimulant. Lockheed did a bang-up job. Their work sailed through the command chain without a hitch.

NF-104
AEROSPACE RESEARCH PILOT SCHOOL

Three F-104s were fitted with the J-2 engines and X-15 reaction control systems. Except for a short-lived disagreement with Lockheed about proposed cost overruns, the modifications were completed without serious problems.........and without cost overruns. Initial flight tests were flown by Jack Woodman, a Lockheed test pilot. The highest altitude reached during flight test was 120,800 ft. Not too shabby for an F-104.

Training experiences with the rocket powered F-104 were not too good. A Navy student bailed out of one. Chuck Yeager survived a bail out of a second one. He did, however, suffer some severe burns in the process. The third modified F-104 is mounted on a pedestal in front of the USAF Test Pilot School.

ARMED FORCES STAFF COLLEGE

Shortly after graduating with Class One, Katie, the boys, and I moved to Norfolk, Va, where I attended Armed Forces Staff College. Pilot students flew Navy TV-2s at Oceana Naval Air Station, for the six months we were there.

I received a telephone call from Frank Borman while we were at Norfolk requesting recommendation letters for him and Jim McDivitt to support their applications for astronaut duty with NASA. Of course, they made the NASA team with no problems.

AEROSPACE RESEARCH PILOT SCHOOL - SECOND TIME

After graduation, I was reassigned back to Edwards AFB. When I reported in at Edwards, I was assigned to the Aerospace Research Pilot School as Deputy Commandant. The School really didn't need a Deputy Commandant. But it was fun to be back in the high-performance aircraft. During the six months that I was away, the number of F-104Bs assigned to the school was increased. In addition, the school acquired a variable stability F-101 and an F-106B. The F-104As and F-104Cs were used for zoom maneuvers and low lift-to-drag ratio landings. The F-104Bs were used for dual instruction and for low lift-to-drag ratio landings.

The variable stability F-101 exposed pilots to a range of instabilities. Instability level one was not much of a challenge. Instability level two was a challenge. At instability level three, the pilot lost control and triggered the release button quickly.

Before going to Armed Forces Staff College, I had planned to teach at the Air force Academy. So, after about six months at Edwards, I contacted Colonel Gibson at the Air Force Academy to see if they were still interested in having me teach there. Col. Gibson was a senior officer on General Davis' staff at Holloman Air Force Base when we were assigned there. He was an exceptional officer and a real gentleman. Col. Gibson started the wheels turning; and, in a matter of a few weeks, I had orders assigning me to the Astronautics Department at the Air Force Academy.

F-101

While Katie, the Boys and I were packing for the move to Colorado Springs, promotion orders arrived at Edwards promoting me to Lieutenant Colonel below the eligibility zone. Naturally, I was very pleased. But the promotion did make things a bit awkward. I would be filling a captain's slot in the Academy Astronautics Department.

Just before we were to depart Edwards for Colorado Springs, "Fox" Stephens, a seasoned test pilot who I had known at Holloman Air Force Base, came by to see me. He invited me to fly on a program to which he was assigned at Edwards as a test pilot. When I asked him what he was flying, he said that he couldn't tell me, but that it was a very interesting airplane. Since I had asked Col. Gibson for help getting to the Academy, I decided it would not be right to back out now. So, I declined the offer. Later, I discovered that he was talking about the SR-71. From a flying standpoint, going to the Academy was not the best choice I ever made. On the other hand, God ultimately determines what we do.

Flying at the Air Force Academy was a long way down the priority list of things to do. The faculty pilots were pretty much limited to the number of hours required to meet the annual minimum flying requirement. All flying was done in T-33s at Lowry Air Force Base near Denver.

Near the end of the first semester at the Air Force Academy, I started looking around for an opportunity to move into a slot that called for a lieutenant colonel. Surprisingly, I quickly discovered a Group Commander job that would be open the first of the year. Colonel McDermott (soon to be General McDermott), the head of the academic half of the Academy, gave me a hearing and agreed to release me to the military side of the house.

The personnel assignment change had to go to Officer Personnel in the Pentagon for concurrence. The Pentagon response was not supportive. In a nutshell, their reply said, "If you can't use an engineering type lieutenant colonel in a technical slot at the Air Force Academy, we have just the place for him here in the Pentagon. Enclosed are his transfer orders." Sometimes, managing your assignments doesn't work exactly as planned.

Flying was done at Andrews Air Force Base. There were two types of aircraft available. One could be assigned to fly C-47s (Gooney Birds) or T-39s. Naturally, all jet pilots were competing for a slot in T-39 Flight. As things turned out, I wound up in C-47s.

GOONEY BIRD

My only flight, after completing the checkout, was a weekend cross flight to Fayetteville, Arkansas to visit Terry at the University of Arkansas. The next week, I was reassigned to T-39s. Looking back, I am very happy that I checked out in the C-47. This aircraft has a long and colorful history that I can appreciate better after flying it.

The T-39 was an excellent flying machine. The best way to describe the T-39 is to refer to it as a twin engine, cargo F-86. It flew well on one engine at GCA approach speeds with the gear down.

T-39

16

Beyond the Air Force

RETIREMENT AND CAREER MOVE TO NASA

W hen I opted not to compete for astronaut duty with NASA in 1959, I told myself that I would never join NASA in any other capacity. That would make me a "ground-pounder" in a flying outfit. Two happenings in the Spring of 1965 led me to reverse that decision:

1. Shortly after I arrived at the Pentagon, the Air Force adopted a policy which grounded pilots who were forty-three years old and who had been a rated pilot for twenty-three years. I was both of these things. Grounded officers would continue to receive flying pay for the rest of their career. But they would not fly military airplanes.

2. Out of the "blue", Deke Slayton called me one day about moving to NASA   Headquarters as an Air Force detailee. A NASA Headquarters Operations Office was in the process of being created. My duty assignment would be Chief of Flight Crew Operations in this office. From the position description, it was clearly a staff job with no management responsibilities. That's what I had in the Pentagon. I told Deke that I wasn't interested.

Shortly after Deke's first call, he contacted me again. This time he asked if was interested in retiring from the Air Force and joining NASA as a civil servant. Ordinarily, I would have responded with, "No Way!" However, the upcoming grounding action and four years in the Pentagon as a staff officer was not something I looked forward to. With considerable uncertainty about whether or not I was doing something I would regret, I chose to move across the river to NASA.

In a matter of a few months, the Headquarters Operations Office lost the contest with the NASA Centers for the command role in running space operations. The office continued to exist. But it had been neutered. Fortunately, I was able to move to Johnson Space Center as Deke's Assistant for Skylab Flight Crew Operations. This was followed by duty as Deke's deputy for conducting the Shuttle Approach and Landing Tests at Edwards AFB. Both of these turned out to be very enjoyable assignments.

Even better, I was allowed to fly T-38s as an "Administrative Pilot" for another fifteen years. The T-38 is a terrific aircraft. The "Y" models that we had at the Test Pilot School for a while had poor control harmony. Once that deficiency was corrected, the T-38 had absolutely no bad characteristics. It will go supersonic. It is a good acrobatic machine. And, it has no single-engine problems. I was able to fly the T-38 for over a thousand enjoyable hours during my NASA time.

T-38

TEXAS A & M UNIVERSITY TEACHING DUTY

In 1984 I retired from NASA to join the Aerospace Engineering Faculty at Texas A & M University. Sharing the teaching and learning process with the Aerospace Engineering Department faculty, staff, and students for thirteen years was an unforgettably wonderful experience, for which I will always be grateful.

CIVILIAN FLYING

Early in the '70s I picked up an FAA Instructor Rating, so that I could teach my sons to fly. This began a general-aviation flying hobby that has steadily grown, making possible sixty-one continuous years of flying.

I thank GOD for my good fortune. In 1942, I didn't expect to live through the war. There is absolutely no doubt in my mind that I have been carried most of the way.

17
Images

Left to Right: Crew Chief, Pilot, Armorer

No more Flights for a Great Flying Machine

Examining an Abandoned Macchi 202 at Catania Main Airfield

Firing a 20 mm Cannon from an ME-109 Left on the Catania Airfield

LEFT TO RIGHT STANDING

Lieutenants: Paul Waters, Mike Adams, Uriel Alford, Eugene Cunneely, Jack Blom-
gren, William Knuttel, Gerald McIlmoyle, Gibson, Robert McDermott, Arthur Lusty,
R.N. Blackburn, Leo McNulty, John Doherty, John Haisty, H.E. Stripling, A.V. Chapin,
John Slocum, Hall Daniel, George Worth, Bob Mays, Bob Stafford, C.R. Swift

LEFT TO RIGHT KNEELING

Major J.K. Scrivner, Captain Tom McElmurry, Major George Halliwell, Major Jack
Pope

One of the Two F-100s Configured for Sidewinder Launches

Preflight for Sidewinder Demonstration at the 1958 World Wide Gunnery Meet –
Nellis AFB

China Lake Sidewinder Test Pilots (3 Navy, 1 Air Force, and 1 Marine Corp)

USAF Test Pilot School Class 56-B Graduates and Faculty

Gathering of Part of the Holloman Sidewinder Project Team
1 Sgt. Wiley (F-104 Crew Chief), 2 Maj. Robin Hansen (Boss), 3 Capt. Dick Corbett,
4 Gearld Strome (Sidewinder Tech Rep), 5 Capt. Ed Moore, 6 Capt. Carl Wheaton

Did Something Right!

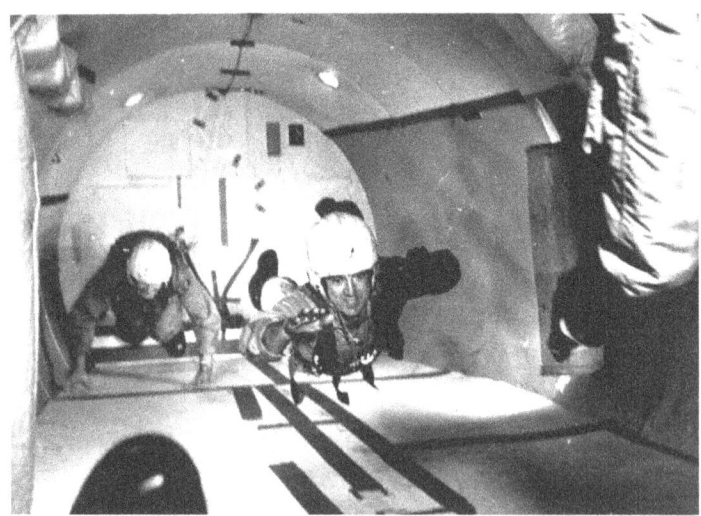

ARPS Class I Zero "g" Indoctrination USAF Zero "g" Aircraft

Upon Arrival at Palmsdale Flight Test Center, First Aerospace Trainer (LAC S/N 1050) is Prepared for Complete Testing of All Systems and Electrical Circuits

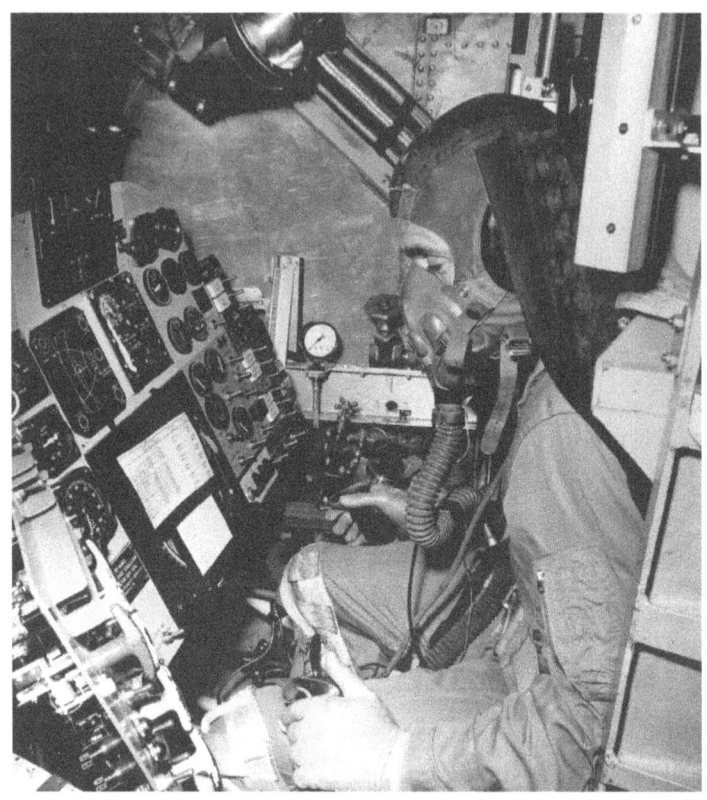

ARPS Class I Performing Mercury Re-Entry Trask at Nine "Gs"

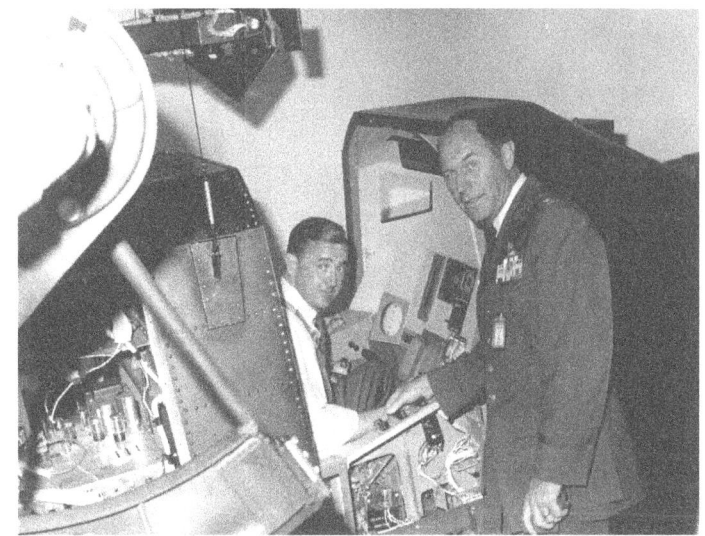

ARPS Class I Trying the Lockheed NF-104A Simulation

ARPS Class I Wired for Data Tracking, Johnsville Centrifuge

ARPS Class I Graduation December 1961

United States Air Force Aero Space Research Pilot School

Review

If you enjoyed this book, will you please do us the great honor of leaving a review on your platform of choice? Your review means so much to us! It is value feedback to us, and it helps other readers decide if this is a book they too would enjoy reading.

Thank you in advance for taking time out of your very busy and important day to leave a review! We appreciate you!

Obituary

Thomas Uriel McElmurry, Lieutenant Colonel, USAF, Ret. passed away on November 3. 2006 at the age of 85.

Tom began his Air Force career at the age of 17 with the Army Air Force in Seward Alaska, eventually receiving flight training in Santa Maria California in 1942. After flying two tours of duty in World War II, he obtained a bachelor of science in mechanical engineering, University of Alabama and master of science in aeronautical engineering, University of Michigan.

Remaining on activity military and as a test pilot, Tom was Chief of the Operations and Training Branch, USAF Experimental Flight Test Pilot School, Edwards AFB, and an Astronaut candidate NASA group 2, selected in 1961 for military astronaut class 1; Instructor, USAF Aerospace Research Pilot School where he was involved with the NF104 space plane; test pilot at the Naval Weapons Center in California and at Holloman Air Force Base in New Mexico.

During his career with NASA, he was Chief, Flight Crew Operations, Office of Manned Space Flight (OMSF); Chief, Orbiter Atmospheric Flight Test Office; Director of Flight Crew Operations; Deputy Manager, Operations Integration, Space Shuttle Program; Office Manager, Operations Integration, Space Shuttle Program Office; Visiting Lecturer, Embry-Riddle Aeronautical University, Prescott, Arizona; Assistant for Systems Reliability and Quality Assurance, Systems Division, Mission Operations.

Upon retiring from NASA, he served 13 years as a Visiting Associate Professor, Aerospace Engineering Department, Texas A&M University. While at A&M he was the advisor to the student Aviation Association as well as to the Corp of Cadets. In his spare time, Tom was a very active flight instructor and avid flyer. With all his accomplishments and adventures, he made it very clear that the only important thing in his life was his relationship with God. He shared his faith with all he met and exemplified his Christian life in how he lived each day.

Bio

EDUCATIONAL BACKGROUND

B.S. in Mechanical Engineering, University of Alabama, Tuscaloosa, Alabama, 1940

M.S. in Engineering, Air Force Institute of Technology, Wright Patterson Air Force Base, Dayton, Ohio

MILITARY EXPERIENCE

Lieutenant Colonel, United States Air Force (Retired)

MILITARY CAREER

United States Air Force (USAF) Experimental Flight Test Pilot School, Edwards Air Force Base (AFB), Edwards, California (1956)

Chief of the Operations and Training Branch, USAF Experimental Flight Test Pilot School, Edwards AFB, Edwards, California (1959-1960)

USAF Aerospace Research Pilot School, Edwards AFB, Edwards, California (1960-61) Instructor, USAF Aerospace Research Pilot School, Edwards AFB, Edwards, California (1961-1962)

NASA CAREER

NASA Headquarters, Washington, D.C. Chief, Flight Crew Operations, Office of Manned Space Flight (OMSF) (1964-1967)

NASA Manned Spacecraft Center/ Johnson Spacecraft Center, Houston, Texas Technical Assistant for Advanced Planning, Director of Flight Crew Operations (1967-1973)

Chief, Orbiter Atmospheric Flight Test Office, Director of Flight Crew Operations (1974-1978)

Deputy Manager, Operations Integration, Space Shuttle Program Office (1978-1980)

Manager, Operations Integration, Space Shuttle Program Office (1980-1982)

Visiting Lecturer, Embry-Riddle Aeronautical University, Prescott, Arizona (Summer 1983)

Assistant for Systems Reliability and Quality Assurance (SR&QA), Systems Division, Mission Operations (1982-1984)

POST-NASA CAREER

Visiting Associate Professor, Aerospace Engineering Department, Texas A&M University, College Station, Texas (1980s-early 2000s)

PROFESSIONAL & HONORARY SOCIETIES

Member, Mu Chapter of Theta Tau (Co-ed Engineering Fraternity)

SELECT PUBLICATIONS

McElmurry, Thomas U. "Aerospace Design Education at Texas A&M University." AIAA Paper 88-4413. AIAA, AHS, and ASEE, Aircraft Design, Systems and Operations Meeting. Atlanta, Georgia, 1988.

SELECT BIOGRAPHICAL REFERENCES

Borman, Frank, Countdown: An Autobiography, With Robert J. Serling (New York: Silver Arrow Books, 1988), 79.

Slayton, Donald K. "Deke," Deke! U.S. Manned Space: From Mercury to the Shuttle, With Michael Cassutt, (New York: Tom Doherty Associates, 1994), 70-71.

Slayton, Donald K, Director of Crew Operations, 25 August 1967, "New Staff Member," Skylab Series Chronological Files, Assistant Director for Flight Crew Operations July 1967 – December 1967, Box 504, History Collection, Scientific and Technical Information Center, Lyndon B. Johnson Space Center, Houston, TX.

Telephone Directories, Manned Spacecraft Center and Johnson Space Center, 1967-1984, Scientific Technical and Information Center, Lyndon B. Johnson Space Center, Houston, Texas.

NASA Thomas McElmurry Oral History Transcript https://historycollection.jsc.nasa.gov/JSCHistoryPortal/history/oral_histories/McElmurryTU/TUM_8-23-00-amended.pdf

United States Air Force Accident Reports 1911-1955 (search "McElmurry" in Pilot's Name field.) https://aviationarchaeology.com/dbSearchAF55.asp

United States Air Force Accident Reports 1956-TODAY (search "McElmurry" in Pilot's Name field.) https://aviationarchaeology.com/dbSearchAF56.asp

Marty Racine. "Life in the air the only way to go for pilot." *Houston Chronicle,* February 3, 2002. https://www.chron.com/life/article/Life-in-the-air-the-only -way-to-go-for-pilot-2067991.php

VISUAL REFERENCES

Thomas McElmurry USAF Pilot Rank: Command Pilot

Thomas McElmurry USAF Officer Rank: Lieutenant Colonel

Found a Typo?

Found a typo? Email christine@jungworks.comwith the subject line of "Typo."

I've tried my very best to make sure this copy is perfect, but I'm only human! If you found an error, do let me know. I want to make sure the next reader has the very best possible version of this book. Thank you, in advance, for taking the time to submit an email to contribute to making this book even better!

Also By Jung Works

Finish the Race | The Great Mountain Adventure

Four cars embark on an exciting race through the mountains when they encounter unexpected challenges along the way. Who will win? Will they even be able to finish the race?

A fun children's book for ages 0-6. Written and illustrated by Thomas C. Jung.

Purchase the eBook on Amazon, or visit jungworks.com for other purchasing options as well as other formats (print, hardcopy, audiobook, ebook, signed author copies.)

Finish the Race | Super Cars Origin Story

The adventure continues! The race cars have just encountered a magic gas pump. Will they recieve special powers?

A fun children's book for ages 0-6. Written and illustrated by Thomas C. Jung.

Purchase the eBook on Amazon, or visit jungworks.com for other purchasing options as well as other formats (print, hardcopy, audiobook, ebook, signed author copies.)

1926 | American Scenes

1926 | American Scenes is available in eBook, paperback, hardcover, and audiobook. To learn how to purchase, please visit jungworks.com.

Earlier Days Flying

Earlier Days Flying is available in eBook, paperback, hardcover, and audiobook. To learn how to purchase, please visit jungworks.com.

Bulk Discount

Interested in buying 10 or more copies? Email christine@jungworks.com with the subject line "Bulk Discount" for more information.

About Jung Works

Jung Works is the umbrella for **Christine Jung, Thomas C. Jung**, & **Ethan Jung**'s projects. Three books, a screenplay, & new art are in the works! We invite you to join us on our creative journey.

Jung Works... A Creative Place

jungworks.com

About Christine Jung

The editor and publisher of this book, Christine Jung, is Thomas McElmurry's granddaughter.

Christine is a writer, equestrian, classical pianist, and artist. Her days start with a cup of coffee and end with a glass of wine (preferably enjoyed outdoors.) In addition to being momager to a young actor, Christine keeps busy with her full-time job, six animals, and whatever her latest project might be. She is inspired by the books she reads, the places she travels and motivated by her supportive husband and son.

jungworks.com

Acknowledgments

MANY THANKS!!

TO THE UNITED STATES AIR FORCE MUSEUM FOR USE OF AIR-CRAFT PHOTOGRAPHS FROM THE MUSEUM'S OUTSTANDING COLLECTION.

-Thomas U. McElmurry 2003

www.ingramcontent.com/pod-product-compliance
Lightning Source LLC
Chambersburg PA
CBHW051615120626
46551CB00014B/1811